SpringerBriefs in Earth System Sciences

Series editors

Gerrit Lohmann, Bremen, Germany
Lawrence A. Mysak, Montreal, Canada
Justus Notholt, Bremen, Germany
Jorge Rabassa, Ushuaia, Argentina
Vikram Unnithan, Bremen, Germany

More information about this series at http://www.springer.com/series/10032

Ionut Cristi Nicu

Hydrogeomorphic Risk Analysis Affecting Chalcolithic Archaeological Sites from Valea Oii (Bahlui) Watershed, Northeastern Romania

An Interdisciplinary Approach

 Springer

Ionut Cristi Nicu
Interdisciplinary Research Department—
    Field Science
Alexandru Ioan Cuza University of Iasi
Iasi
Romania

ISSN 2191-589X   ISSN 2191-5903 (electronic)
SpringerBriefs in Earth System Sciences
ISBN 978-3-319-25707-5   ISBN 978-3-319-25709-9 (eBook)
DOI 10.1007/978-3-319-25709-9

Library of Congress Control Number: 2015952047

Springer Cham Heidelberg New York Dordrecht London

Springer International Publishing AG Switzerland is part of Springer Science+Business Media
(www.springer.com)

Foreword

I am delighted to write a foreword for the book *Hydrogeomorphic Risk Analysis Affecting Chalcolithic Archaeological Sites from Valea Oii (Bahlui) Watershed, Northeastern Romania. An Interdisciplinary Approach*. This book presents the results of the doctoral research undertaken by Ionut Cristi Nicu. The focus of this research is the risks that natural phenomena can present to Chalcolithic archaeological sites in north-eastern Romania. In addition, this book discusses other topics related to the dynamics of past populations in a small catchment area and to interactions between humans and their changing environments.

The research presented in this book provides critical new data that contributes new understandings of Romania's past. Although archaeological excavations have taken place in Romania for more than a century, questions remain regarding the placement of settlements, treatment of the dead and the mobility of the populations. This book helps to answer some of these questions.

This study presents practical and up-to-date case studies from a landscape with great archaeological potential, the Cucuteni-Trypillia culture from Eastern Europe. The book brings something new and refreshing to global archaeology, as it treats endangered sites and the possible future directions in preserving these sites.

During my tenure as president of the World Archaeological Congress from 2003–2014, I become very aware of the cultural and natural risks that are posed to cultural heritage sites worldwide. As with other countries, the protection and preservation of archaeological heritage in Romania is a critical issue. A particular challenge in Romania is the development of a complete registry of existing sites and recording the continuous degradation caused by the active processes of natural erosion and anthropic activity as well as from looting activities. However, notable work has been undertakenrecently in identifying and adding as many sites as possible to the National Archaeological Registry (RAN) managed by the Ministry of Culture, and the official database of archaeological heritage developed by the Institute of Cultural Memory (CIMEC) and the National Heritage Institute (INP). Since 2000 and 2013 RAN has listed around 30,000 archaeological entities and close to 20,000 sites some from more than 5,500 localities. Nevertheless, Romania is still far from having a complete database of archaeological heritage. Research concerning the unintended effects of natural risks is growing throughout the world, in line with considerable endeavours being undertaken in rescue archaeology and in cultural heritage preservation. While the Romanian scientific literature in this area is still sparse, with just a limited number of applicative studies, young scholars such as Ionut Cristi Nicu are making important contributions.

I am pleased to recommend this inter-disciplinary book to scholars in the areas of landscape archaeology, natural hazards, cultural heritage, archaeological geographic information systems, physical geography, human-environment interactions and geoarchaeology. I look forward to more books being produced in this growing area of scholarship.

September 2015   Prof. Dr. Claire Smith
                                                   Head, Department of Archaeology,
                                                    Flinders University, Australia

Acknowledgments

This work was supported by the Partnership in Priority Domains project PN-II-PT-PCCA-2013-4-2234 no. 314/2014 of the Romanian National Research Council, Non-destructive approaches to complex archaeological sites. An integrated applied research model for cultural heritage management and from a Marie Curie International Research Staff Exchange Scheme Fellowship within the seventh European Community Framework Program (acronym: FLUMEN, Project no. 318969, FP7-PEOPLE-2012-IRSES). The Laboratory of Geoarchaeology from Alexandru Ioan Cuza University of Iasi is kindly acknowledged for the logistic support. Many thanks to my colleagues from the Interdisciplinary Research Department—Field Science for helping me during the field trips and last but not the least to my family for the moral and sometimes financial support.

Contents

Introduction

The book was written as a result of the author's Ph.D. thesis, while being a novelty and having a need for geographical and archaeological literature in Romania. The author originally a geographer, working in an academic environment since finishing his bachelor studies. Working alongside and supporting archaeologists and geographers alike, while managing to combine the two subjects in a particular manner. The ability and technical skills needed to analyse each component of the environment, in close relation with Chalcolithic archaeological sites and integrating in GIS is achieved through commitment, hard work, practice and a good understanding of the principles involved. The interdisciplinary character of the work is appreciated in Romanian academic circles already.

The study area chosen represents the place of discovery of Cucuteni culture, which took its name from *Cetățuia* settlement and is considered to be one of the oldest in Europe, which is of high significance for Romanian archaeology, as well as for European and global archaeology.

The degradation of archaeological heritage through erosion is the main focus of the work, as the title indicates. The three case studies are representative and treat current problems facing the world heritage (case study of Băiceni-Cucuteni gully is inspired mainly because it affects an archaeological site of high importance and has the potential to affect another site). Soil degradation processes are increasingly active across the globe. Recent decades have shown that meteorological and hydrological processes have increased this risk.

Moldavian Plateau located in North-eastern Romania is strongly affected by gully erosion, landslides and sheet erosion. The most important causes that contribute to these processes are caused by deforestation, a high friable substrate (clay, sand, marl, and loess), torrential rains and inadequate application of land improvement works. By monitoring and analysing the degradation process, important data is saved and anti-erosion and conservation measures are proposed to local authorities and stakeholders and in some cases implemented, in their attempt to mitigate the effects of geomorphological processes on archaeological sites and human settlements. This example is also highlighted in the last chapter of the book.

The entire work is unique as a way of treating a subject of archaeology in interdisciplinary context. The table of contents was realised as a novelty in the field. The instruments used for surveying by the author are of the latest technology. Interpretation and results obtained are of high interest at an interdisciplinary level. The book can be useful both for Geography and Archaeology students and also those who follow courses as: Geoarchaeology, Landscape Archaeology, and GIS in Archaeology, Natural Risks, Heritage Management, and Heritage Studies. Likewise, the book is easily readable and comprehensive to non-specialised readers.

A book like this brings new ideas and substance in analysing the environment and development of human society, in connection with the most importance resource, water.

Prof. Dr. Gheorghe Romanescu

Chapter 1
Geographic Framework

Abstract The study area is located in the north-eastern part of Romania, with a surface of 97 km², at the intersection of the parallel of 47°21′0.86″ lat. N with the meridian 26°49′37.07″ long. E and the intersection of parallel of 47° 13′23.32″ lat. N with the meridian of 27° 10′35.68″ long. E. The catchment has a length of 31 km, a maximum altitude of 443.19 m and minimum altitude of 61.57 m. It represents the area of discovering of Cucuteni culture; moreover, it has been continuously inhabited since prehistoric times, being known in the Romanian geomorphological literature as «Poarta Târgului Frumos» (Târgul Frumos Crossing Point).

Keywords Bahluieț · Iași county · Moldavian plain · Northeastern Romania · Pastoral movement

1.1 Regional Settings

Valea Oii is located in the northeastern part of Romania and occupies a central-western position in the Bahlui watershed (Fig. 1.1). After the digitization of the topographical plans (scale 1:5000, edition 1979), it derived a catchment area of 97 km². Emerging from the spring until it flows into the Bahluieț River in the Sârca locality, the main course has been given different names, depending on where the settlements are along the river. From the spring until Băiceni village, it is called Pârâul Rece (Cold Creek), Pârâul Oaia (Sheep Creek) to Mădârjești dam. From Mădârjești dam outlet until the entry into Sârca Lake and the confluence with the Bahluieț brook, it is called Trestiana.

From a mathematical perspective, the intersection of the parallel of 47°21′ 0.86″ lat. N with the meridian 26°49′ 37.07″ long. E marks the northernmost point in the basin, Stroiești Hill (444 m) and the intersection of the parallel of 47°13′ 23.32″ lat. N with the meridian of 27°10′ 35.68″ long. E marks the southernmost point of the basin, at the spill of Oii creek in the Bahluieț River, Sârca locality. The eastern

© The Author(s) 2016
I.C. Nicu, *Hydrogeomorphic Risk Analysis Affecting Chalcolithic Archaeological Sites from Valea Oii (Bahlui) Watershed, Northeastern Romania*,
SpringerBriefs in Earth System Sciences, DOI 10.1007/978-3-319-25709-9_1

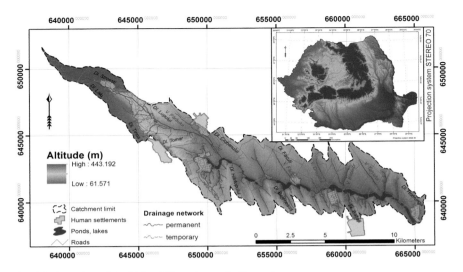

Fig. 1.1 Location of the Valea Oii catchment in Romania

boundary is the intersection of the parallel of 47°14′ 33.11″ lat. N with the meridian of 27°11′ 25.26″ long. E.

The study area is bordered on the north and northeast with Măgura catchment (74.21 km²), Putina catchment (16.28 km²), Bahlui catchment (2023 km²), northwest with Bâdiliţa catchment (22 km²), west with Hărmăneşti (41 km²), south with Păşcănia catchment (11.32 km²), Probota catchment (10 km²), Cucuteni catchment (12.45 km²) and finally Bahluieţ catchment (110.56 km²). The basin has a length of 31 km, a maximum width of 6.5 km (direction N–S, to the east of the village of Boureni) between Ciobanului Hill and Jora Hill. The middle and lower part of the basin has a minimum width of 2–3 km. The maximum altitude is 443.19 m and the minimum is 61.57 m, which means there is a 381.62 m difference in the level (Nicu and Romanescu 2015).

1.2 Geographical Regionalization and Administrative Division

Although it has a small area, the basin extends over two well individualised areas in the Moldavian Plateau: Moldavian Plain (occupying 90.5 % of the total area of the basin) and Suceava Plateau (an area of only 9.5 %). From the administrative point of view, Sheep Valley basin lies exclusively in the Iaşi county; it includes the following communes Todireşti (7126 ha total area), Cucuteni (2826 ha) Baltati (4508 ha), Belceşti (10,390 ha), Cotnari (10,335 ha) with the following villages: Stroeşti, Băiceni, Cucuteni, Balş, Boureni, Filiaşi, Podişu, Gugea, Valea Oii, Bălţaţi and Sârca. This administrative fragmentation of the basin is not beneficial. The location of the six communes surrounds the basin, causing less attention from local authorities on land improvement works taking place within the area.

In the past, numerous populations of people congregated in this area (Romanians from Muntenia, Transylvania, Bukovina, Saxons, Csangos, Gypsies, Lipovans) moving from the mountains to the lower plains, which were rich in natural resources (very fertile soils suitable for agriculture). The area has been continuously inhabited since prehistoric times, and known in the literature as ≪Poarta Târgului Frumos≫. This long transitional period has left its mark in the local toponymy. Villages of Gypsies: Ruginoasa, Heleşteni, Miclăuşeni. Traces of the Saxon origin: Hărmăneşti—prefix Hărman = German equivalent of Hermann (Tufescu 1941). Of a particular importance, remains a pastoral movement within the land, over time information has remained in the toponymy *oaia* (sheep), which gave its name to the village Valea Oilor (Sheep Valley) and also to the main river. Sheep breeding represents the major occupation even to the present day.

References

Nicu IC, Romanescu G (2015) Effect of natural risk factors upon the evolution of Chalcolithic human settlements in northeastern Romania (Valea Oii watershed). From ancient times dynamics to present day degradation. Z Geomorphol. doi:http://dx.doi.org/10.1127/zfg/2015/0174. (in press)

Tufescu V (1941) O regiune de vie circulaţie: "Poarta Târgului-Frumos". BSRRG, LIX, M. O., Imprimeria Naţională, Bucureşti

Chapter 2
Methodology and Research Techniques

Abstract In order to accomplish the proposed results, a series of modern research techniques were used. The DTM was generated by digitizing topographic plans scale 1:5000 (1979 edition), which was basically the base layer for other geospatial and archaeological data. Other maps from different years and scale were integrated in the study. Digital data was combined with data from the fieldtrips, GPS and geomorphological surveys. Finally, all the spatial data (geographical and archaeological) was integrated in a GIS database and, in this way, a high diversity of spatial analysis was possible (proximity analysis, viewshed analysis, 3D visualisation, etc).

Keywords Georeference · GPS · STEREO 70 · Topographic map · Total station

2.1 In Situ Identification of the GCPs (Ground Control Points) of Different Orders and Their Distribution in the Catchment

Following the analysis of the map that contains the distribution of the GCPs in the basin, it can be observed that a large amount of the area has been marked and surveyed. The presence of a total of 12 GCPs that cover a territory of 97 km^2 indicates that the undertaking of topographical measurements is easily covered (being that it is performed with a total station or the geodetic GPS). Thus, there are 2 GCPs of 2nd order, 3 GCPs of 3rd order, 6 GCPs of 4th order and one GCP of 5th order (Bălțați Orthodox Church) that it is just a reference and back sight ID for the measurements done with the total station (Fig. 2.1).

© The Author(s) 2016
I.C. Nicu, *Hydrogeomorphic Risk Analysis Affecting Chalcolithic Archaeological Sites from Valea Oii (Bahlui) Watershed, Northeastern Romania,*
SpringerBriefs in Earth System Sciences, DOI 10.1007/978-3-319-25709-9_2

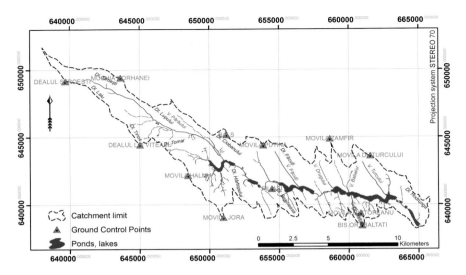

Fig. 2.1 Distribution of GCPs in the Valea Oii catchment

2.2   Utilisation of Old and New Maps (Austrian Maps, Historical Army Maps)

In order to obtain a scientific approach, a varied cartographic background with a different representation scale will be used. 1:50,000 (1894 edition), scale 1:25,000 (1984 edition), topographic plans scale 1:5000 (1979 edition), historical army maps scale 1:20,000 (1942–1945 edition), topographic plans scale 1:2000 (1973 edition) but also orthophotomaps (taken during the flights of 2005 and 2008), satellite images.

T period 1970–1993, was realized the topographical plan of Romania with about 90 % of the countries territory, as follows: scale 1:2000 on 12 %, scale 1:5000 on 75 % (thus explaining the lack of 9 topographic plans from the study area), scale 1:10,000 for 3 % (Mihăilă et al. 1995). Using a detailed cartographic background, allows the realisation of complex analysis, both in terms of natural elements, evolution, and for economic transactions.

2.2.1   Methodological Staging

In the fulfilment of the geographical database, GIS software was used: TNTmips, AutoCAD 2008, Global Mapper, and ArcGIS. In the execution of the DTM, the topographical plans at a 1:5000 scale was used, with the following stages: collection and selection of the cartographic support from the ANCPI (National Agency for Cadastre and Land Registration) Iasi, scanning, importing in GIS, georeferencing

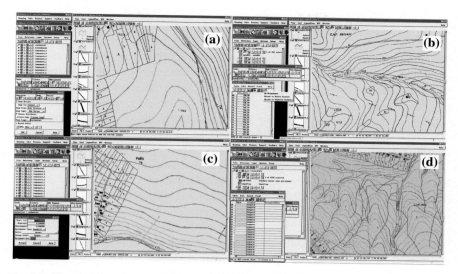

Fig. 2.2 The workflow for obtaining a DTM. **a** Contour lines digitisation. **b** Adding altitudinal points. **c** Assigning Z value for contour lines. **d** Drainage network classification

raster cartographic objects in STEREO 70 projection system (for a precise overlay of the topographic surveys over the topographic plans), manual digitisation (Fig. 2.2a); in order to obtain a more accurate modelling of the relief, altitudinal points have been added (Fig. 2.2b), numerical classification by assigning geographic information to objects (Fig. 2.2c, d). STEREO 70 cartographic system is based on the same mathematic principles established and applied for the old projection system from 1930. Besides, this projection system belongs to the ones which distorts radial lengths, but still retains the values of the angles (Mihăilă et al. 1995).

In the upper basin, where topographic plans are missing tracks of the area were taken, using GPS Leica RTK Rover 1200. The tracks were then embedded in the same file with the altitudinal points from the topographic plans. Out of several methods of contour lines and altitudinal point's interpolation (Kriging, Topo to Raster, IDW), was chosen Topo to Raster method, being the most suitable for the present case, resulting in a DTM with 5 × 5 m pixel size. It can be observed a faithful reproduction and high quality of the reality in the field, and can be used for both analysis of the location of the known settlements and the location of new archaeological sites on the basis of the characteristics of the terrain (Fry et al. 2004). It can be combined successfully with orthophotoplans, satellite imagery, LANDSAT images, and 3D views (Katsianis et al. 2008).

In the last years, modelling and 3D reconstruction are used more often in the capitalisation and presentation of research results in geography, and also in archaeology. From habitat reconstruction, the evolution of relief and reconstruction of objects, 3D modelling constitutes an indispensable tool in the interdisciplinary research.

2.2.2   Database Integration in GIS

Ghilardi and Desruelles (2008) and Ghilardi et al. (2008) shows the importance of integrating a geographical database (connected to the mapping and conversion of the in situ data in digital format, realisation of spatial analysis, 3D visualisation, analysis through statistic methods of the distribution of settlements, paleogeographic reconstructions, all offering a better interpretation of the relationships of sites, their structure and their formation), with an archaeological one (placement of sites) in a GIS.

Within the 1990s, archaeological studies were based only on a 2D perspective. With the advancement of new technologies, which improved the quality and efficiency of the digital analysis (satellite images, aerial photographs along with in situ observations, detailed topographical measurements), the transition toward a tridimensional study took place. With the help of this tool, a multitude of surveying (visibility, proximity, etc.) could lead to solving the proposed premises and to better understand the archaeological sites placement in certain locations, in accordance with certain factors derived from GIS analysis (altitude of relief, soil type, and proximity toward water resources).

References

Fry GLA, Skar B, Jerpåsen G, Bakkestuen V, Erikstad L (2004) Locating archaeological sites in the landscape: a hierarchical approach based on landscape indicators. Landscape Urban Plan 67 (1–4):97–107. doi:10.1016/S0169-2046(03)00031-8

Ghilardi M, Desruelles S (2008) Geoarchaeology: where human, social and earth sciences meet with technology. SAPI EN S 1(2):1–9

Ghilardi M, Fouache E, Queyrel F, Syridres G, Vouvalidis K, Kunesch S, Styllas M, Stiros S (2008) Human occupation and geomorphological evolution of the Thessaloniki Plain (Greece) since Mid Holocene. J Archaeol Sci 35(1):111–125. doi:10.1016/j.jas.2007.02.017

Katsianis M, Tsipidis S, Kotsakis K, Kousoulakou A (2008) A 3D digital workflow for archaeological intra-site. J Archaeol Sci 35(3):655–667. doi:10.1016/j.jas.2007.06.002

Mihăilă M, Corcodel G, Chirilov I (1995) Cadastrul general şi publicitatea imobiliară – bazele şi lucrările componente. Editura Ceres, Bucureşti

Chapter 3
Geological Characterisation

Abstract The area overlaps the Moldavian Platform. The dominant deposits in the catchment belong to Bassarabian. Bassarabian sediments are mainly composed of an alternation of marls, sands, clays, with a significant thickness of about 1000 m, especially in the northern part of Cucuteni village and west of Baiceni village. The general pitch of the strata is on NNW-SSE direction, which lends the relief a characteristic morphology with extended structural surfaces, and front of cuestas. The superficial deposits belong to the Quaternary and are present as thick loess deposits. As a consequence of the geological deposits, soil erosion processes are very frequent in the eastern part of the country.

Keywords Bessarabian · Moldavian Platform · Mollusc · Oolithic sandstones · Volhinian

3.1 Petrography and Structural–Tectonic Aspects

From a geological point of view, the Valea Oii watershed entirely overlaps the Moldavian Platform (Fig. 3.1). A paper of great importance for this study is the one of Ştefan (1989), where the results of in situ research are underlined between the years 1978–1988; with a major emphasis on the evolution of Sarmatian mollusc, and on the separation of Sarmatian stacks which outcrops the region in many lithologic units, with mineral resources being extremely utile (oolithic limestone, sandstones, sands, clays and mineral waters), with an economic importance at a local level. As from a stratigraphic point of view, the limit between Volhinian and Bessarabian was set based on the faunal criteria. The making of the geographical map of the Dealul Mare Hârlau area is at the 1:50,000 scale. Some sections are of main interest for the understanding and placement of human settlements in this area.

Bessarabian deposits dominate the basin (clay marls with sand intrusions), and in the lower half Pleistocene terrace deposits have been encountered (sands and gravels). The domination of Bessarabian deposits is known for having fine

I.C. Nicu, *Hydrogeomorphic Risk Analysis Affecting Chalcolithic Archaeological Sites from Valea Oii (Bahlui) Watershed, Northeastern Romania,*
SpringerBriefs in Earth System Sciences, DOI 10.1007/978-3-319-25709-9_3

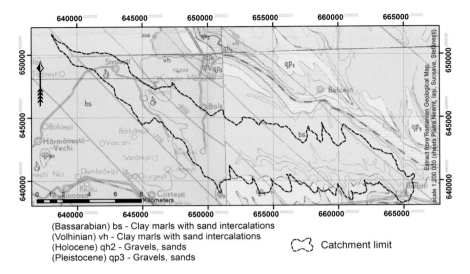

(Bassarabian) bs - Clay marls with sand intercalations
(Volhinian) vh - Clay marls with sand intercalations
(Holocene) qh2 - Gravels, sands
(Pleistocene) qp3 - Gravels, sands

Catchment limit

Fig. 3.1 General geological map of the study area

granulation and high friability; this is the main factor for the current geomorphological processes with a high frequency throughout the basin, especially on the right
side. The upper part of the basin had been studied with the main focus regarding the
geological characteristics, natural resources, and their exploitation. The clay was
used by prehistoric people in the construction of their houses and the making of
vessels.

In the upper side of the basin, at about 1.5 km northwest from Băiceni–Cucuteni
gully, there is a natural reserve of national interest of palaeontology (coordinates
47°17′ 52.23″ N, 26°54′ 48.02″ E). This natural palaeontological reserve highlights
the lithological deposits that occur at the base in the following order: Băiceni clays
and sands, Bahlui Sirețel sands, Hârlău oolith, Sticlăria-Sângeap sands, Crivești
oolithic sandstones (2014). The purpose of the reserve is to protect the brackish
mollusc, characteristic to the early Bessarabian: *Mactra pallasi Bayli*, *Tapes gre-
garius ponderosus d'Orb*, and *Gibula podolica insperata Kol*. This fauna is very
well preserved, with entire robust and well-ornamated valve shells, which can be
found in extremely high numbers. Alongside these forms, emerge other sarmatian
taxons.

Reference

Ștefan PD (1989) Geologia regiunii Dealul Mare-Hârlău și perspectivele în resurse minerale utile.
    Univ. "Al.I.Cuza" Iași

Chapter 4
Relief

Abstract Based on the DTM different analysis of the relief is made (hypsometry, slope, aspect), as well as the geomorphological map of the catchment. The genetic types of relief are analyzed in detail. An important part of this chapter is the analysis of placement of archaeological sites based on morphology and morphometry. The analysis has highlighted that the prehistoric populations preferring to place their settlements on relatively low altitudes 100–200 m and gentle slopes. The viewshed analysis made from all archaeological settlements is showing that the prehistoric populations had a very good visibility (over 80 %) and a very good control of the surrounding territory.

Keywords Alluvial deposits · Sculptural plain · Quaternary · Structural plateau · Terrace

4.1 Morphographic and Morphometric Characterisation

Whereas the geomorphologic processes that are taking place in this basin are only remembered or studied in a broader sense, some case studies are imposed. The Moldavian Plain is totally overlapped on the Moldavian Platform with a long paleogeomorphologic evolution (over 70 million years) which is also in a state of present evolution (Băcăuanu 1967).

The left-side watershed of the basin coincides with the limit between the Bahluieț basin and the one of Bahlui and partially with the Măgura, outlining itself through: Dl. Stroiești (444 m), Muchia Corhanei/Dealul Osoi (365 m), Dl. La Rupturi (251 m), Dl. Măgurii (238 m), Dl. Țărna (187 m), Dl. Ciobanului (189 m), Dl. Lipoveanului (198 m), Movila Putinii (192 m), Dl. Făcuți (190 m), Dl. Polieni (160 m), Movila Zamfir (177 m), Dl. Basnea (150 m), Dl. Turcului (172 m) and Coasta lui Donici (162 m).

The right slope outlines the Bahluietului basin limit through: Dl. Stroiești (420 m), Dl. Cepei (385 m), Dl. Laiu (375 m), Dl. Tinos (345 m), Dealul lui

© The Author(s) 2016
I.C. Nicu, *Hydrogeomorphic Risk Analysis Affecting Chalcolithic Archaeological Sites from Valea Oii (Bahlui) Watershed, Northeastern Romania*,
SpringerBriefs in Earth System Sciences, DOI 10.1007/978-3-319-25709-9_4

Viteazul (340 m), Dl. Halmu (256 m), Movila Jora (189 m), Dl. Hârtopului (183 m), Dealul Bejeneasa (183 m), Dl. Boghiului (169 m), Dl. Bălțați (152 m), Movila Hârtopeanu (176 m), Dl. Mândra (137 m) and Dl. Sârca (126 m).

The network of the valleys existed almost entirely at the end of Pliocene and the beginning of Quaternary; the rivers were having the same general direction, with the exception of some small slides attributed to subsequent valleys. The subsequent character has been imprinted also in the Valea Oii basin; almost, the cross profile of the entire length of the valley is asymmetric; the right slope being steep and the left one slightly slanting. Another aspect imprinted in the relief on the south limit of the basin, covered with colluvial deposits which constitutes the terrace of the Bahluieț basin. The alluvial material carried along the valley has been submitted under the form of alluvial deposits, which due to external modelling factors have evolved in time under the form of local terraces, scattered on the left slope, uphill from Făcuți village or on the slopes of Bejeneasa Hill (Băcăuanu and Martiniuc 1966).

Taking into consideration the altitudinal difference of the basin, for accomplishing a hypsometric map (Fig. 4.1), the fallowing altitude classes were chosen: 61.5–100, 100.1–150 m, 150.1–200 m, 200.1–250 m, 250.1–300 m, 300.1–350 m, 350.1–400 m și > 400.1 m. The areas which do not succeed 100 m altitude represent 12.4 % of the total surface being spread from the Bahluietului confluence until the half of the basin, where there are placed the main accumulations (Sârca, Mădârjești, Dobre, Ichim, Podișu); also, the terraces of the two slopes are included with a greater share for the left one. Altitudes between 100.1 and 150 m get the upper side of the basin with 41.9 % occupying the inferior third of the slopes on the left side; the majority of the right slope has the upper hand in the middle basin; the stage between 150.1 and 200 m holds 26.6 % of the total, it met on the extremities

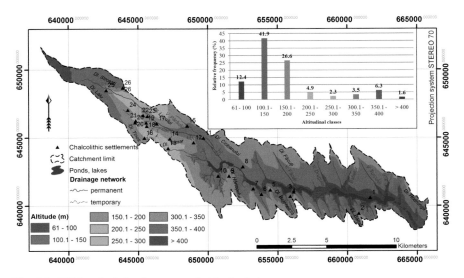

Fig. 4.1 DTM and relative frequency of altitudinal classes

of both slopes from the effusion and until the borders of the Boureni and Bals villages, where the transition towards the plateau area takes place, that is accomplished through the 200 and 300 m stages (7.2 %). The hypsometric class of 300.1 and 350 m with a 3.5 % share, the one of 350.1–400 m (6.3 %) and the one of >400 m (1.6 %) include the surfaces that are existing solely in the Dealul Mare Hârlău in the NNW extremity also known as the Broscăria Laiu plateau.

4.2 Genetic Types of Relief (Structural, Sculptural, of Accumulation)

Within the basin, we find a denudational relief represented through sculptural and structural forms but also an accumulation relief (Fig. 4.2).

4.2.1 Structural Relief

Structural relief is characterised by the presence of cuestas, found on the right side of the basin and throughout the structural plateau from the upper part of the basin Laiu. If in the Moldavian Plane two cuestas can be set apart, colluvial cuesta and valley flank cuesta, in our study basin only the delluvial cuestas are present, almost entirely formed out of the slope processes releases, developed mainly on deposits of loess, clay and loam.

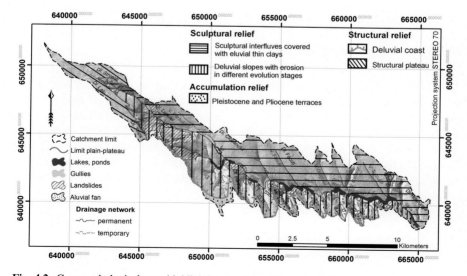

Fig. 4.2 Geomorphological map highlighting the three types of relief

4.2.2 Sculptural Relief

It occupies the biggest surface of the basin on both of the slopes. During the formation of this type of relief, the main morphogenetic role is held by the external factors represented through the hydrographic network, the sum of the slope processes to which the climatic conditions are added and the presence of the soft rock sedimentation complex. Within this type of relief, we find the following:

– *sculptural interstream areas* covered in eluvial clay with light washing processes which are met on the left side of the basin, with linear slopes with a tilt of no more than 3–5°, the peaks evolution and stream plateau is due to some weak alteration processes, degradation and erosion, the descent of the general surface of the relief is slow, through the means of some wide wavy formes, the possibility of natural conservation of the soil and even the formation of a 3–4 cm slim counterpane of loessoidal clay. Through time, these streams were analogized with erosion platforms. The sculptural streams undergo the form of hills and low plateaus (Dl. Lupului, Dl. Ciobanului, Dl. Făcuți).
– *colluvial slopes* with mixed degradations in multiple staged of evolution, spread on the right side of the basin, where the slopes exceed 3° leaning. Here the majority of the surface erosion processes take place (gutters, trenches, gullies, torrents) favored by the Sarmatian clay substrate; a special type of relief is the one developed on saline Sarmatian deposits or when, due to intense evapotranspiration and the low groundwater level, the salt reaches the surface soil through capillarity; to these fine texture saline deposits, washings are characteristic. Landslides with a wide diffusion are acquiring feature to this landscape, the majority being stabilized landslides.

4.2.3 Accumulation Relief

Is represented by the Pleistocene and Pliocene terraces met in the inferior half of the basin, but also in plains, terraces, alluvial cones. The plains were formed in the postglaciar period through the succession of erosion and accumulation periods, with 3–20 m thickness and are occupying the lowest portions of the relief. (Băcăuanu 1967). A very good example is from the upper basin, between Lupului Hill and Mănăstirii Hill, where both of the versants are affected by sliding processes, resulting a typical accumulation relief, where currently is located Băiceni village.

4.3   Location of Archaeological Sites Based on Morphology and Morphometry

The utilisation of a high quality and resolution cartographic stock, in the present case the DTM with 5 × 5 m/pixel resolution, can help in obtaining more precise results. Internationally, for obtaining the DTM there are different methods and techniques used, with a direct applicability over archaeological sites: satellite images from different years (CORONA—Goosens et al. 2006; Casana and Cothren 2008; ASTER, SPOT, LANDSAT), maps and topographic plans at different scales (Parmegiani and Poscolieri 2003), aerial photography, direct measurements in the field with the GPS and total station, 3D laser scanner (Balzani et al. 2004), LiDAR (Harmon et al. 2006; Coluzzi et al. 2010; Schindling and Gibbes 2014), all these methods successfully integrated in GIS (Wescott and Brandon 2005; Harrower 2010), where different analysis (e.g. viewshed) can be made. The last one, LiDAR, is a very precise method, but also very expensive and its usage is precise to just some high interest areas.

From Fig. 4.3, referring to the classification of Chalcolithic archaeological sites on altitudinal classes, there can be observed the fact that a number of 17 sites (the majority belonging to the Precucuteni and Cucuteni phases) are placed on the 100–200 m altitudinal difference (difference which occupies more than half of the basins surface), spread in the middle and superior basin until the contact with the plateau area, the preference for lower and relatively high formes is evident, where the slopes were permitting it, agriculture was practiced (Asăndulesei 2012), but also the natural protection of the settlements.

It is not the case of the *Dealul Mândra* (no. 1) settlement, found at an altitude of 73 m, in the proximity of the main course of the valley, the inhabitants could

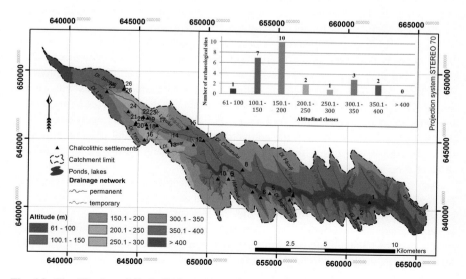

Fig. 4.3 Classification of Chalcolithic settlements on altitudinal classes

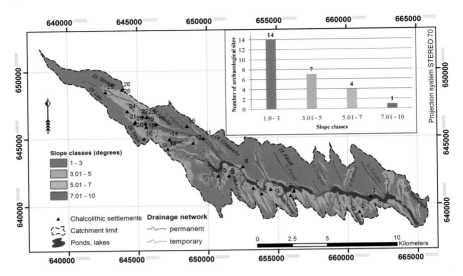

Fig. 4.4 Classification of Chalcolithic settlements on slope classes

practice agriculture on the south slope. Proof of the fact that the settlement was located without taking in consideration other factors, such as natural defence, is that it exists only one archaeological layer—Cucuteni A, this settlement being abandoned at the end of this period.

The transition through the higher plateau area, with altitudes between 300 and 400 m, where there are met a number of 5 sites, is accomplished almost suddenly, and in the 200–300 m class there are a number of 3 sites. In particular, the settlements on higher altitudes were holding an essential role, the one of defence, inside these settlements fortification systems being found.

The mobilities of the Chalcolithic populations cannot be completely understood, if the slope is not taken into consideration for the placement of settlements, their defence, but also terrestrial relationships, all of these with a minimum of physical effort (Fig. 4.4). The preference for setting up the settlements in places with gentle slopes (3–5°), with a number of 21 sites is obvious, all the other sites being met in the slope class between 5 and 7° (4 sites), and also a site in the slope class 7–10°. The last ones are met either at the contact between the plain and plateau, either on the right side of the basin, on the front of cuesta presently affected by the sheet erosion processes.

An important analysis widely used in archaeology is viewshed analysis (Wheatley 1995; Ellis 2004; Kim et al. 2004; Rua et al. 2013). For this study area, a cumulative viewshed was made for the 26 settlements (Fig. 4.5). As it can be observed, the visible surface covers over 80 % of the total catchment surface, which means that the Chalcolithic populations had a very good control of the surrounding territory. This analysis can help us identify the dynamics of the population inside the catchment.

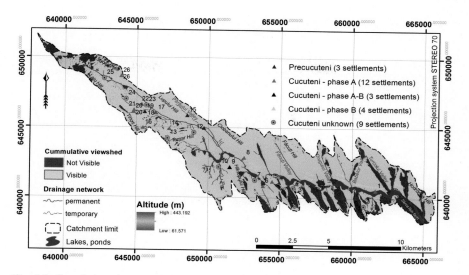

Fig. 4.5 Cumulative viewshed for the Chalcolithic settlements

References

Asăndulesei A (2012) Aplicații ale metodelor geografice și geofizice în' cercetarea interdisciplinară a așezărilor cucuteniene din Moldova. Studii de caz. PhD Dissertation, Univ. "Al. I. Cuza" Iași

Băcăuanu V (1967) Cercetări geomorfologice asupra Câmpiei Moldovei (extras). An. șt. ale Univ. "Al. I. Cuza" (serie nouă), secț. II, tom XIII, Iași

Băcăuanu V, Martiniuc C (1966) Cercetări geomorfologice asupra teraselor din bazinul Bahluiului. An. șt. ale Univ. "Al. I. Cuza", secț. II, tom XII:147–156

Balzani M, Santopuoli N, Grieco A, Zaltron N (2004) Laser scanner 3D survey in archaeological field: the forum of Pompeii. In: International conference on remote sensing archaeology, Beijing, pp 169–175

Casana J, Cothren J (2008) Stereo analysis, DEM extraction and orthorectification of CORONA satellite imagery: archaeological applications from the Near East. Antiquity 00:1–18

Coluzzi R, Lanorte A, Lasaponara R (2010) On the LiDAR contribution for landscape archaeology and palaeoenvironmental studies: the case study of Bosco dell'Incontrata (Southern Italy). Adv Geosci 24:125–132. doi:10.5194/adgeo-24-125-2010

Ellis SJR (2004) The distribution of bars at Pompeii: archaeological, spatial and viewshed analyses. J Rom Archaeol 17:371–384

Goosens R, De Wulf A, Bourgeois J, Gheyle W, Willems T (2006) Satellite imagery and archaeology: the example of CORONA in the Altai Mountains. J Archaeol Sci 33:745–755. doi:10.1016/j.jas.2005.10.010

Harmon JM, Leone MP, Prince SD, Snyder M (2006) LIDAR for archaeological landscape analysis: a case study of two eighteenth-century Maryland plantation sites. Am Antiq 71 (4):649–670

Harrower MJ (2010) Geographic information system (GIS) hydrological modeling in archaeology: an example from the origins of irrigation in Southwest Arabia (Yemen). J Archaeol Sci. doi:10. 1016/j.jas.2010.01.004

Kim Y-H, Rana S, Wise S (2004) Exploring multiple viewshed analysis using terrain features and optimisation techniques. Comput Geosci 30(9–10):1019–1032. doi:10.1016/j.cageo.2004.07. 008

Parmegiani N, Poscolieri M (2003) DEM data processing for a landscape archaeology analysis (Lake Sevan—Armenia). In: The international archives of the photogrammetry, remote sensing and spatial information sciences, vol XXXIV, part 5/W12, pp 255–258

Rua H, Gonçalves AB, Figueiredo R (2013) Assessment of the lines of Torres Vedras defensive system with visibility analysis 40(4):2113–2123. doi:10.1016/j.jas.2012.12.012

Schindling J, Gibbes C (2014) LiDAR as a tool for archaeological research: a case study. Archaeol Anthropol Sci 6(4):411–423. doi:10.1007/s12520-014-0178-3

Wescott KL, Brandon RJ (eds) (2005) Practical applications of GIS for archaeologists. A predictive modelling toolkit. Taylor & Francis, London

Wheatley D (1995) Cumulative viewshed analysis: a GIS-based method for investigating intervisibility, and its archaeological application. Archaeology and geographical information systems. In: Stančič Z (ed) Lock G. Taylor & Francis, London

Chapter 5
Hydrography

Abstract Analyzing water resources from a certain area can provide us important information regarding the spread and dynamics of prehistoric population. Within the study area the present water resources dynamics (ponds) are analyzed based on old topographic maps; the hydrological data available made possible the analysis of extreme hydrological events, such as flooding. The area has a very good protection against floods, in present being a number of ten ponds. An important analysis is represented by the proximity of archaeological sites towards a water resource (river course and tributaries and springs).

Keywords Coast springs · Dam · Floods · Mineral waters · Ponds

Romania is ranked twenty first in Europe for water resources (Gâştescu 2010). Valea Oii catchment (cadastral code XIII.1.15.32.12.7) (Atlasul Cadastrului Apelor din România 1968) enframes itself in the category of the first degree from the Prut basin, a sub-basin of Bahlui river. Botoşani and Iaşi counties hold second and third places for the acquatic surfaces (after Tulcea county) (Gâştescu 2010).

The hydrological factor has a particularly important role in shaping the landscape, modelling began with successive withdrawal of Sarmatic Sea, being in constant evolution even today. Within the catchment are mentioned sulphurous-bicarbonate-sodium springs from Băiceni area, which were formed during the Middle Sarmatian (Macarovici and Bejan 1957). Water resources are indicators of possible evolution of prehistoric populations; water is the most important resource for socio-economic activities, development of new habitation areas and life in general. Water characteristics have been observed and used since ancient times (sulphurous waters, waters with a high content of salts, mineral waters).

5.1 Hydrogeological Characteristics

The study area falls in the general hydrogeological features of Moldavian Plateau. It can be distinguished, hydrogeologically speaking, into two processes: accumulation and storage of groundwater in Sarmatian rocks (with slightly brackish taste) and

I.C. Nicu, *Hydrogeomorphic Risk Analysis Affecting Chalcolithic Archaeological Sites from Valea Oii (Bahlui) Watershed, Northeastern Romania,* SpringerBriefs in Earth System Sciences, DOI 10.1007/978-3-319-25709-9_5

confined waters in terrace deposits (with slightly sweet taste) (Pantazică 1974; Minea 2012). Around the catchment there are hydrogeological stations of first order at Mădârjeşti, Belceşti and Cotnari, and second order at Ruginoasa. Groundwater oscillation is closely related with the rock properties where it is stored; aquifer was intercepted at −1.22 m depth at F1—Mădârjeşti drilling. In the upper basin, inside Băiceni-Cucuteni gully, the hydrostatic level was located using GPR technology at 8–10 m depth (Nicu and Romanescu 2011).

Geotechnical studies have been conducted on the current locations of dams, providing important data regarding the substrate, the deepest reaching −12 to −13 m. These drills were aimed at identifying strata and checking their thickness (through particle size analysis) to determine the ideal site for the location and construction of dams in a safe and sustainable way. The dam was constructed at this location because the physico-mechanical properties of the soil, resulting from geotechnical profiles and grain size analysis, have allowed this (yellow clay with a high degree of impermeability).

5.2 Lake Basins Morpho-Hydrographical Evolution

The existence and study of lacustrine accumulations within a given territory represents a high importance from an economic and landscape point of view. The necessity of the existence of these accumulations (lakes, ponds), especially in the Moldavian plain, is due to the presence of the continental temperate climate with prolonged drought and running waters with a low flow (Băican 1970). The permanent stream is due to a high supply of underground water 40–60 % (Romanescu et al. 2008).

Within this basin, the stagnant waters are represented by the lacustrine units, with a number of 10 ponds (Boureni, Boureni fish pond, Bejeneasa, Filiaşi, Podişu, Ichim, Dobre, Mădârjeşti, Sârca, Sârca fish pond), of an anthropogenic origin. The level and evolution of these waters is variable, due to factors such as changing seasons, meteorological conditions (precipitations), from how the ponds utilised (simple-attenuation of the flash floods, complex-attenuation of the flash floods, pisciculture, recreation), but also from the geomorphological conditions (which sometimes has lead to the disappearance of some ponds due to landslides). Based on Water Management Headquarters from Iaşi decision no. 106/28.06.1960, embankment work and dam constructions took place. The execution period of the dams was between 1961 and 1962, the year of commissioning being 1962.

The presence of ponds in the north-eastern part of the country was recorded in several cartographic documents, of which the oldest dates from 1600, entitled "Fishing and fish farming in Romanian regions during the upper feudalism—1600", the map of Moldova (Bawr) surveyed between 1768 and 1774 (scale 1:308,000). Overall, lake basins morphohydrographical evolution in Moldova, starting from fifteenth century, is attributed to the economic development of the country during seventeenth to nineteenth centuries, and increasing the number of people (Băican 1970). Most of the ponds were built in the Moldavian Plain.

To highlight the lake basins morphohydrographical evolution, the following topographical maps and plans were used: topographic maps scale 1:50,000 (1894 edition); historical army maps scale 1:20,000 (1945 edition); topographic maps scale 1:25,000 (1984 edition); orthophotomaps scale 1:5000 (2005 edition). At the same time, the data from Water Cadaster Atlas and S.C. Piscicola Podişu S.R.L. Administration were consulted.

After digitising the maps and plans resulted a set of four maps, from which can be traced the evolutionary trend of the lake basins. In year 1968, the total volume of water in the lakes was 4.281 million m^3, with an average flow of 115 l/s (Atlasul Cadastrului Apelor din România 1968). Currently, the total volume of water stored within the catchment is 25.754 million m^3, of which Sârca pond has the highest volume 23.300 million m^3 (90.4 % of total, Table 5.1).

Initially, ponds were used for animal consumption (especially sheep), rarely for irrigation (Minea 2005). After analysing the maps by Băican (1970), it can be observed, for the period of seventeenth to nineteenth centuries, in the upper part of the basin a number of five ponds, of which only one is present to this day; while downstream there are three ponds. For the period 1801–1850 (Fig. 5.1), the ponds around Băiceni village disappeared, the total number of ponds in the basin being six, located downstream from present day Filiaşi village.

Table 5.1 Ponds situation from study area

No. pond	Name	Administrator	Total volume (million m^3)	Dam height (m)	Importance class	Associated risk value
1	Bejeneasa	S.C. Piscicola Podişu S.R.L.	0.090	4.0	IV	0.21
2	Boureni	S.C. Piscicola Podişu S.R.L.	0.042	4.0	IV	0.21
3	Dobre	S.C. Mihpes S.R.L.	0.448	5.0	IV	0.14
4	Făcuţi/Podişu	S.C. Piscicola Podişu S.R.L.	0.378	6.0	IV	0.17
5	Filiaşi Pepinieră	S.C. Piscicola Podişu S.R.L.	0.176	5.0	IV	0.19
6	Filiaşi	S.C. Piscicola Podişu S.R.L.	0.163	5.0	IV	0.21
7	Ichim	S.C. Piscicola Podişu S.R.L.	0.264	6.0		
8	Mădârjeşti	S.C. Mihpes S.R.L.	0.574	5.0	IV	0.12
9	Sârca	D.A. Prut-Bârlad	23.300	17.0	II	0.25
10	Sârca— Pepinieră	S.C. Mihpes S.R.L.	0.319	5.0	IV	0.24

Source Administraţia Bazinală de Apă Prut-Bârlad (http://www.rowater.ro/daprut/default.aspx. Accessed on 02 June 2015)

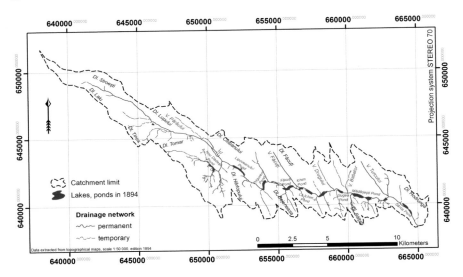

Fig. 5.1 The limit of the ponds in 1894

When changing the landuse from pastures to cropland, during 1950s–1960s (collectivisation period, Fig. 5.2), and after in the 1980s (Fig. 5.3) a higher quantity of water for irrigation (as needed for an area of about 300–400 ha) was necessary. However, with the promulgation of Land Law (Law nr. 18/1991) and therefore agricultural land entering under private property, the functionality of lakes has changed, being used for fish farming. Currently, irrigation is made only on very

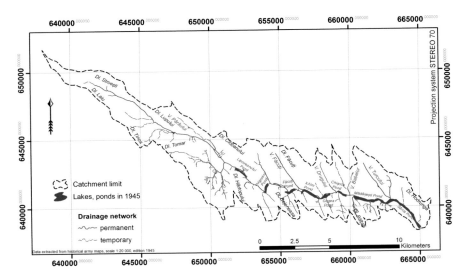

Fig. 5.2 The limit of the ponds in 1945

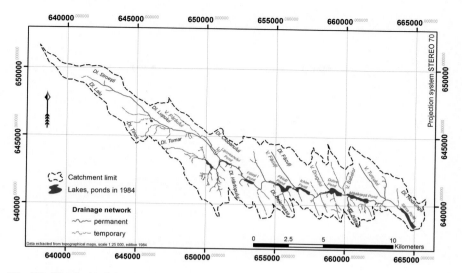

Fig. 5.3 The limit of the ponds in 1984

limited areas, with facilities that run on fuel. Boureni pond, Boureni fish farm, Bejeneasa, Filiaşi, Podişu and Ichim ponds are under the administration of S.C. Piscicola Podişu S.R.L. intended for fish farms during the summer.

Genesis and evolution of lake basins differs from an area to another. Within this catchment (as in the whole area of Moldavian Plain), except fluvial erosion, a key role is held by the geomorphological processes (landslides) (Ujvari 1959). In this case, the landslide is responsible for the disappearance of Lipoveanul pond (Nicu and Romanescu 2015); which lead upstream of the landslide to the newly built Bejeneasa pond in 1962 (with the dam height of 4 m and a total volume of 0.090 million m³). Human interventions also had an impact on the evolution of lake basins, by the draining of ponds and increasing agricultural areas (having hydro-morphic soils with high fertility). Between 1945 and 1984 it can be observed the disappearance of Prigoreni pond (located upstream of Filiaşi village) and Gugea pond (located near Gugea village).

Over time, the general trend was to increase the surface of ponds (Fig. 5.4) from upstream to downstream, ponds with small areas being transformed into agricultural land, all these transformations were having a high demand on water supplies for fish farms, irrigations, and also the socio-economic development of the area. In recent years, global warming has had a negative impact with average precipitations below 550 mm and high evapotranspiration (Minea 2004), and as a consequence Bejeneasa pond has dried up; what is now left is a dense reed of vegetation. Additionally, the role of ponds is being that of defense against floods (Minea 2012). Overall, the evolution of lake basins within the catchment was and still is subject to various factors: natural (landslides) or anthropogenic (drainage, construction of new dams).

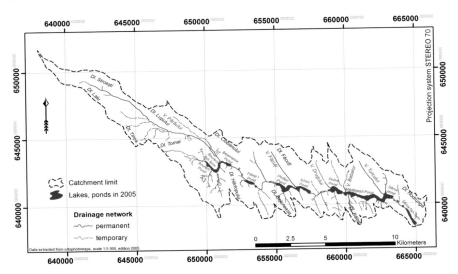

Fig. 5.4 The limit of the ponds in 2005

Another approach is the one of the origin of pond names (hydronims) being able to give us information on: issues relating to ethnic origin of populations (Lipoveanul pond—slavic population widespread especially in the area of Târgu Frumos village), or their occupations (Lippovans are known for intensive cultivation of vegetables); another example is Prigoreni pond, a Prigorean meaning a person dealing with beekeeping.

5.3 Hydrological Risks (Floods)

The flood represents the sudden rise in a short time of the level and water flow of a river, over the usual values. As a result, neighbouring territories are temporarily or permanently covered with water. Other names accepted are overflow or downpour (Romanescu 2009). They are formed under the specific climatic conditions of hilly area, when a maximum amount of precipitation falls in 24 h (>100 l/m^2), sometimes accompanied by sudden melting of snow; another important element is the low degree of afforestation of this area. In general, the highest amounts of precipitations in 24 h occur during the summer months (June–July) (Romanescu et al. 2012).

The formation and propagation of floods have an important role in the morphohydrographic characteristics of the basin (Table 5.2); these are calculated based on the topographic maps. Besides the possibilities for computing and prevention of hydrological risks associated with maximum overflow, the knowledge of the parameters are of importance both in analysis and evolution of hydrological regime, and the local management of water resources (Minea and Romanescu 2007). The

Table 5.2   Morphohydrographic parameters of the catchment

Nr. crt.	Valea Oii river basin	Value
1	Maximum length (L_{max})	33.5 km
2	Average length (L_{med})	14.9 km
3	Average width (l_{med})	6.5 km
4	Surface (S)	97 km^2
5	Shape factor (F_f)[a]	0.08
6	Circularity report (R_c)[b]	0.78
7	Asymmetry coefficient (a)	0.02

[a]Proposed by Horton (1932)
[b]Proposed by Miller (1956)

indices shape factor (F_f) and circularity report (R_c) with values 0.08, respectively, 0.78 (values under 1), indicates that the basin has an elongated shape. Asymmetry coefficient (a) indicates that the area of the two versants is approximately equal (47.8 km^2 for the right side of the basin and 49.2 km^2 for the left side of the basin), water resources being equally distributed.

For Bahlui river basin, between 1950 and 2006, there were no fewer than 62 floods (Minea 2012). Floods took place in the summer of 1975, when the nine dams within the catchment were broken; the amount of water from Valea Oii was, however, retained by Podu Iloaiei dam (about 80 %), which would have flooded Iaşi city with devastating consequences (Pantazică 1974).

The documentation of the designing and construction of dams from the valley recorded a flood from June 1975; the average precipitation values recorded at the three meteorological stations around the catchment are: 212.3 mm at Podu Iloaiei, 180.6 mm at Târgu Frumos, 167.7 mm at Cotnari, and 178.4 mm at Strunga (Fig. 5.5). Calculations have been made for the mitigation of flows (1, 5 %, Table 5.3) for each dam, as well as the maximum flow occurence for the safety levels.

All the dams within the basin are classified as class IV of importance, the possible overflow could be easily controlled, as long as these values are not exceeded. The dam with the most important role is Sârca, located downstream of all other ponds, with the role of mitigating floods for Iaşi city. The annual average flow of Valea Oii river in Sârca pond is 0.165 m^3/s, flow determined by correlation with the hydrological station from Podu Iloaiei where direct measurements were made. Calculations for the maximum values of flow and volume were made, without taking into account the existence of the ponds upstream. Sârca dam is designed for a probability of 0.01 %. For any flood with a value higher than 1 %, the dam risks breaking, all the water volume from the nine ponds can be detained by this dam; Sârca pond is part of Iaşi city management scheme against floods.

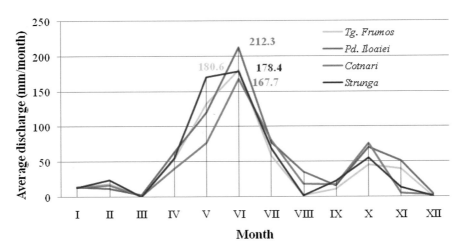

Fig. 5.5 Average discharge associated for flood from June 1975 (data from four meteorological stations around the catchment)

Table 5.3 Mitigated effects for maximum flow occurence with 1 and 5 % probabilities for the ponds from Valea Oii

No. crt.	Pond name	Maximum flow 1 %	Mitigated flow 1 %	Maximum flow 5 %	Mitigated flow 5 %
1	Boureni	89	69	60	47
2	Bejeneasa	98	58	64	38
3	Filiași I	108	45	66	28
4	Filiași II	117	40	70	26
5	Făcuți	126	35	73	24
6	Ichim	138	30	78	23
7	Dobre	150	28	84	20
8	Mădârjești	156	24	90	19
9	Sârca	168	–	–	–

5.4   Location of Archaeological Sites Based on Water Resources

Water represents the necessary element of life on Earth. Human settlements, from ancient times, have taken into consideration the close proximity of water and its benefits when settling in a certain area (Boghian 2004).

In the determination of *distance to water* index, considerations were given for the permanent and temporary water course (extracted from the topographic plans 1:5000 scale), as there was leakage when precipitations took place. Prehistoric populations could have taken advantage of this, being no major changes in the

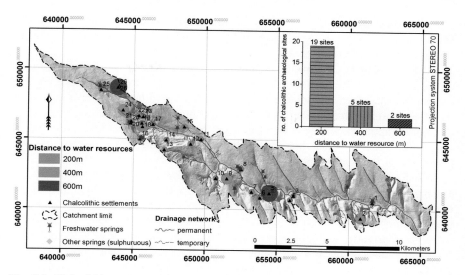

Fig. 5.6 Chalcolithic settlements proximity towards a water resource

structure of the valley network since the end of Pliocene to the beginning of the Quaternary period (Băcăuanu 1967).

For the calculation of this index, the *Ring Buffer* function of the ArcGIS software was used. The distances of the calculation for this index are 200, 400 and 600 m respectively; the maximum value of 600 m was chosen as it represents the longest distance from a site (*Pietrărie*) (no. 26) to the nearest water source. The ponds which currently exist on the main course of the valley have been eliminated from the analysis.

This indicator was calculated for Bahlui catchment, and each of the three phases of the Cucuteni culture, which highlighted medium values of the distance between settlements and the one closest to the water source, as follows: for phase A—401 m, phase A–B—408 m, phase B—414 m (Asăndulesei 2012). Following the completion of the proximity map of archaeological sites towards a water source (Fig. 5.6), from the total of 26 sites, 19 are placed at a 200 m distance from a water source, 5 sites at 400 m distance and only 2 sites at 600 m distance. The average distance from a prehistoric settlement to the nearest water source is 269 m. Being a small basin in comparison with the Bahlui river catchment, the distance of 200 m can be plausible in relation with our basin dimensions. Following this analysis resulted in the close proximity towards a water source constitutes one of the determining factors in the layout of a settlement.

Another aspect which cannot be omitted is the main occupation of the Chalcolithic populations, which was pottery, water being an indispensable element. Also, water was used for nourishment, household utility and sometimes probably the irrigation of crops on small surfaces. A hypothesis issued was one which stated that in case the prehistoric communities did not have easy access to a water resource, they would dug wells for capturing water (the Cucuteni sites from Hăbăşeşti and Truşeşti), even if archaeological evidence has never been found to confirm this fact (Văleanu 2003).

Two special cases are from the archaeological sites *la Iaz/Iazul 3/Dealul Mândra* (no. 1) and *Dealul Boghiu/Dealul* Mare (no. 5). In the first case, the archaeological site no. 1 is placed at an altitude of 73 m, less than 200 m from the water course; at the south-west border of the settlement there are three springs partially clogged (one of them is captured), visible only when Sârca pond level is low (Nicu et al. 2012). In this situation, the fact that the water was the main factor in deciding the placement of this settlement can be stated.

In the second example, *Dealul Boghiu/Dealul Mare* (no. 5) settlement, in the eastern extremity appears a number of two coastal springs, in the middle part of the depletion area of the landslide, which affects the site on the west, north and east part. Having these two springs in the vicinity could have potentially triggered the landslide. Currently, the springs are not captured and do not have a considerable flow, sometimes water formes puddles in the back of the small mounds; these puddles often constitute the place of drinking for numerous sheep in the area. It can be presumed that the prehistoric populations could of used these springs as a source of water, being very accessible, on the east-facing slope, at a distance of approximately 250 m. Stating this fact is due to the difficult terrain to the main course of the valley, being forced to travel a distance of approximately 700–800 m, to go down and up a slope of approximately 9°–10° (being a high altitude settlement 185 m), also making the inhabitants vulnerable in the face of possible enemies.

The adaptation of mankind to the surrounding conditions of the environment, and especially to the presence of water resources, is evident also by the evolution of human settlements which at the beginning were placed in the upper part of the basin; then, with the evolution of the relief and the triggering of the hydrogeomorphological processes, of the floods due to deforestation, began the occupancy of the lower plain areas, close to the main water course, with fertile soils due to the supply from alluvial slopes.

References

Asăndulesei A (2012) Aplicaţii ale metodelor geografice şi geofizice în cercetarea interdisciplinară a aşezărilor cucuteniene din Moldova. Studii de caz. PhD Dissertation, Univ. "Al. I. Cuza" Iaşi

Băcăuanu V (1967) Cercetări geomorfologice asupra Câmpiei Moldovei (extras). An. şt. ale Univ. "Al. I. Cuza" (serie nouă), secţ. II, tom XIII, Iaşi

Băican V (1970) Iazurile din partea de est a României oglindite în documentele istorice şi cartografice din sec. XV – XIX. An. şt . ale Univ. "Al. I. Cuza" XVI:65–73

Boghian D (2004) Comunităţile cucuteniene din bazinul Bahluiului. Editura Universităţii "Ştefan cel Mare" Suceava

Gâştescu P (2010) Resursele de apă din România. Potenţial, calitate, distribuţie teritorială, management. Lucrările Simpozionului Naţional Resursele de apă din România. Vulnerabilitate la presiunile antropice, Editura Transversal, Târgovişte, pp 9–26

Horton RE (1932) Drainage basin characteristics. Eos Trans Am Geophys Union 13:350–361

Macarovici N, Bejan V (1957) Asupra genezei apelor minerale din Moldova, dintre Siret şi Prut. Stud Cerc balneolog – climatolog:259–280, Institutul de Balneologie, Bucureşti

Miller JP (1956) Ephemeral streams. Hydraulic factors and their relation to the drainage net. US geological survey professional paper, vol 262-A, Washington

Minea I (2004) Evaluarea perioadelor secetoase în Câmpia Moldovei. IC.DMP.1:131–142, "Gh. Asachi" Technical University, Editura Performantica, Iaşi

Minea I (2005) Evoluţia unităţilor lacustre din bazinul hidrografic Bahlui. Lucr Sem Geogr "D. Cantemir" 25:127–137

Minea I (2012) Bazinul hidrografic Bahlui. Studiu hidrologic. Editura Univ. "Al.I.Cuza" Iaşi

Minea I, Romanescu G (2007) Hidrologia mediilor continentale. Aplicaţii practice, Casa Editorială Demiurg, Iaşi

Nicu IC, Romanescu G (2011) Determination of ground-water level by using modern non-distructive methods (GPR technology). In: Air and water components of the environment, Cluj-Napoca, pp 441–448

Nicu IC, Romanescu G (2015) Effect of natural risk factors upon the evolution of Chalcolithic human settlements in Northeastern Romania (Valea Oii watershed). From ancient times dynamics to present day degradation. Z Geomorphol. doi:http://dx.doi.org/10.1127/zfg/2015/0174 (in press)

Nicu IC, Asăndulesei A, Brigand R, Cotiugă V, Romanescu G, Boghian D (2012) Integrating geographical and archaeological data in the Romanian Chalcolithic. Case study: Cucuteni settlements from Valea Oii (Sheep Valley—Bahlui) watershed. Geomorphic processes and geoarchaeology. From landscape archaeology to archaeotourism, Moscova—Smolensk, ≪Universum≫, pp 204–207

Pantazică M (1974) Hidrografia Câmpiei Moldovei. Editura Junimea, Iaşi

Romanescu G (2009) Evaluarea riscurilor hidrologice. Editura Terra Nostra, Iaşi

Romanescu G, Romanescu G, Stoleriu CC, Ursu A (2008) Inventarierea şi tipologia zonelor umede şi apelor adânci din Podişul Moldovei. Editura Terra Nostra, Iaşi

Romanescu G, Cotiugă V, Asăndulesei A, Stoleriu C (2012) Use of the 3-D scanner in mapping and monitoring the dynamic degradation of soils: case study of the Cucuteni-Baiceni Gully on the Moldavian Plateau (Romania). Hydrol Earth Syst Sci 16:953–966

Ujvari I (1959) Hidrografia R.P.R. Editura Ştiinţifică, Bucureşti

Văleanu MC (2003) Omul şi mediul natural în neo-eneoliticul din Moldova. Editura Helios, Iaşi

Chapter 6
Climate

Abstract Global climatic changes are affecting all the areas of the world. Especially in the north-eastern part of Romania where, over the last century the average annual temperature increased by 0.5°C. The average annual temperature for the study area is between 8°–9.5°C, and the amounts of rainfall range between 500–700 mm. For this part of the country there are specific heavy rains in a very short timespan, which in addition to the natural and anthropic factors (steep slopes, land use, bad management of land improvement works) accentuates the initiation and development of soil erosion processes that affect archaeological sites.

Keywords Climatic changes · Drought · Foehn · Global warming · Torrential rains

6.1 General View of the Climate

In the Moldavian plain and also in Valea Oii watershed, four main barometric centres are present: Azores anticyclone, Siberian anticyclone, Icelandic cyclone and the Mediterranean cyclones. Slight influences come on behalf of the Scandinavian, Greenland anticyclones and from the north of Africa and the south-western depressions of Asia (Erhan 2004; Mihăilă 2006).

6.2 Temperatures

Thermically speaking, the Moldavian Plain falls into a continental temperate with excessive influences, manifested by frequent periods of drought and torrential rainfall. The average yearly temperature is between 8 and 9.5 °C. A repetitive event that is being encountered at the meteorological station from Cotnari, located in the proximity of Valea Oii upper basin, at 289 m altitude, is recording an average annual temperature of 9.2 °C due to slight air descending phenomena of foehn from

© The Author(s) 2016
I.C. Nicu, *Hydrogeomorphic Risk Analysis Affecting Chalcolithic Archaeological Sites from Valea Oii (Bahlui) Watershed, Northeastern Romania*,
SpringerBriefs in Earth System Sciences, DOI 10.1007/978-3-319-25709-9_6

Suceava Plateau (Erhan 2004). The maximum temperature was 37.6 °C (06.07.1988) at Podu Iloaiei. The minimum temperatures were −24.5 °C at Cotnari (14.01.1972) and −32.3 °C at Podu Iloaiei (20.01.1963). Average annual number of days with frost is 109.8 days, and the number of summer days is 61.5 days at Cotnari meteorological station (Minea 2012).

Climatic changes occured in Europe during the twentieth century which has had a negative impact, (the trend is increasing temperatures especially in spring and summer). This increase proved to be nonlinear and heterogenous at a global scale (Croitoru et al. 2011). For Romania, in the northeast and therefore the present area of study, it is estimated that during 1901–2007 the average annual temperature increased by 0.5 °C, with a higher rate for extracarpathian regions (Hobai 2009). Also, for this part of the country, there were significant increases (during 1961–2010) in air temperatures for the months of June, July and August (Piticar and Ristoiu 2012). Other studies in this area have shown the same trend of air temperature increasing, framing the area in the general trends of global warming.

It has been observed a steady increase of average temperatures, every 10 years, with values ranging from 0.6 to 1.9 °C. Also, synoptic factors dominate over the local ones, the warming trend could be observed in all meteorological stations, whether stations are located at higher altitudes—Ceahlău (1879 m), or in urban areas. Year 1988 was identified as the year in which the changes have started in most meteorological stations from northeast, and a majority of regions in Europe (Beaugrand 2004; Donnelly et al. 2009).

6.3 Precipitations

The geographical placement of the Moldavian plain, east to the Carpathian chain which constitutes a real "orographic barrier" towards the dominant western circulation, determines an uneven distribution of the precipitations quantities. The main consequence of the interference of the Atlantic circulation with the Carpathian chain is the asymmetry of the precipitations quantities between the western part and the eastern or south-eastern ones. A high importance is held by the relegating cyclones formed in the cold season in Baltic Sea region which then travel towards the northwest of the Black Sea, relegating towards east to west and thus affecting the Moldavian plain.

By its geographical position, in the extra-carpathian region, with the influence of continental air masses from the east, northeast and north, Valea Oii catchment receives moderate amounts of rainfall, with values ranging between 500 and 700 mm (Fig. 6.1). Dominant precipitations range between 500 and 600 mm, which are spread over about 80 % of the basin; higher amounts, between 600 and 700 mm, are found only in the transition area from plain to plateau. Distribution of precipitations is 70 % from rains and 30 % from snow. There is a quantitative increase in March–June, when about 75–90 % of the rainfall occurs. The highest quantities fall in the warm season (April 1–September 30). The highest amount of rainfall during the summer, as a result of the torrential character (Mihăilă 2006; Romanescu et al. 2008).

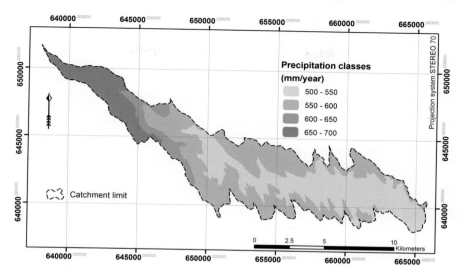

Fig. 6.1 Average precipitation classes

6.4   The Effects of Topoclimatic and Climate Changes on Archaeological Sites

Both climate oscillations and the tendency of temperature rise combined with the torrential rainfall characteristics, have a damaging impact upon the degradation of the archaeological sites. The climatic changes and variations from the past have set their blueprint upon the dynamics of prehistoric populations. These were proven and described by Monah (1985), Boghian (2004). All the archaeological sites from the basin are under the direct affect of the climate variations, three eloquent study cases have been analysed in detail in the final chapter. Climate variations have caused significant degradations in the past and continues to this day, which will inevitably have a negative impact on the future of these archaeological sites. However, an entire ensemble of factors is accelerating this process (land use, the degree of vegetation cover, slopes, bad management of land improvement works and especially the anthropogenic factor).

References

Beaugrand G (2004) The North Sea regime shift: evidences, causes, mechanisms and consequences. Prog Oceanogr 60(2–4):245–262. doi:10.1016/j.pocean.2004.02.018

Boghian D (2004) Comunitățile cucuteniene din bazinul Bahluiului. Editura Universității "Ștefan cel Mare" Suceava

Croitoru AE, Holobaca I-H, Lazar C, Moldovan F, Imbroane A (2011) Air temperature trend and the impact on winter wheat phenology in Romania. Clim Change 111(2):393–410. doi:10. 1007/s10584-011-0133-6

Donnelly A, Cooney T, Jennings E, Buscardo E, Jones M (2009) Response of birds to climatic variability; evidence from the western fringe of Europe. Int J Biometeorol 53(3):211–220. doi:10.1007/s00484-009-0206-7

Erhan E (2004) Aspecte ale foehnizării aerului în estul României. Lucr. Sem. Geogr. "D. Cantemir", pp 23–24

Hobai R (2009) Analysis of air temperature tendency in the upper basin of Bârlad river. Carpath J Earth Env 4(2):75–88

Mihăilă D (2006) Câmpia Moldovei. Editura Universității Suceava, Studiu climatic

Minea I (2012) Bazinul hidrografic Bahlui. Studiu hidrologic. Editura Univ. "Al.I.Cuza", Iaşi

Monah D (1985) Aşezările culturii Cucuteni din România. Iaşi

Piticar A, Ristoiu D (2012) Analysis of air temperature evolution in Northeastern Romania and evidence of warming trend. Carpath J Earth Env 7(4):97–106

Romanescu G, Romanescu G, Stoleriu CC, Ursu A (2008) Inventarierea şi tipologia zonelor umede şi apelor adânci din Podişul Moldovei. Editura Terra Nostra, Iaşi

Chapter 7
Flora and Fauna

Abstract The area falls into the silvosteppe and deciduous forests category of vegetation. Silvosteppe is characteristic for this part of the country. Given the fact that in the past the territory was covered by Sarmatic Sea, there are specific plants developed on a Sarmatian substrate; salty grass has an important economic role within the catchment. Around the ponds there are specific hydrophilic and hygrophilic plants. The only forested areas are located in the upper part of the basin, which currently have a surface of about 200 ha. Forests had a significant role in the development and evolution of prehistoric settlements, wood being one of the most important elements of subsistence.

Keywords Exploitation · Phytocenosis · Forest · Silvosteppe · Vegetation

The current vegetation represents a mixture of natural phytocenosis, partially modified by humans and secondary phytocenosis formed over the old primary ones and then destroyed (Ion et al. 2011). Some vegetation in the area can be considered semi-natural. The vegetation of Moldavian Plain, and of Valea Oii catchment, is determined by specific geographical conditions like: geographical location, geology, relief, climate, hydrography and soils.

Valea Oii catchment, according to the vegetation, falls into the silvosteppe and deciduous forests; from an altitudinal point of view, it also falls into the category of deciduous forests, with two undergrowths (forests of oak and beech) (Aniţei 2000). The silvosteppe from central and northern Moldavia is characteristic for hilly areas between 100 and 300 m altitudes. It appears on chernozems and phaeozems (Ion et al. 2011).

Depending on specific local conditions, some encounters of azonal plant associations such as meadow can be found along the main course of the river. Halophilous vegetation has a close correlation with soils that have a higher content in salts, where they appear on a substrate made of Sarmatian rock, where the level of underground water is shallow. These are found in small areas in the upper basin, around Băiceni-Cucuteni gully, some isolated valleys, and slopes affected by landslides in the middle and lower basin. Salty grass (*Salicornia herbacea* and

I.C. Nicu, *Hydrogeomorphic Risk Analysis Affecting Chalcolithic Archaeological Sites from Valea Oii (Bahlui) Watershed, Northeastern Romania,* SpringerBriefs in Earth System Sciences, DOI 10.1007/978-3-319-25709-9_7

Fig. 7.1   Reed vegetation associations (*Phragmites communis*) on Dobre pond bank (08.11.2010)

Salsoda soda) (Minea 2012), is commonly found and has a high economic importance within the basin; sheepfolds are placed in the proximity of these areas, satisfying the animals need for salt by consuming these plants making it unnecessary for farmers to obtain large lumps of salt.

Ponds, have an important economic role within the catchment; this environment is ideal for the emergence and development of hydrophilic and hygrophilic plants. These are mainly found on the banks of ponds, where the water flow rate is lower, being represented by: reeds (*Phragmites communis*) (Fig. 7.1), sedge (*Carex pseudocyperus*) and bulrush (*Thypha latifolia*).

7.1   The Spread of Forests (Past and Present)

Evidence related to the wooded areas of the country from the past can be found in the testimonies of several historians (L. de Carra 1781; Emil Pop 1941), and maps made over time: *Harta Moldovei* (Map of Moldova) by F.G. Bawr (1769–1772), scale 1:288,000; *Harta rusă* (Russian Map), scale 1:420,000, edition 1835; *Harta topografică a României* (Topographic Map of Romania), scale 1:200,000, edition 1915; *Harta topografică a României* (Topographic Map of Romania), edition 1972–1974, scale 1:200,000. The main cause that led to the reduction of the forest was: expansion of human settlements, in the context of demographic raising, by intensive exploitation for both exports and domestic consumption; as at that time

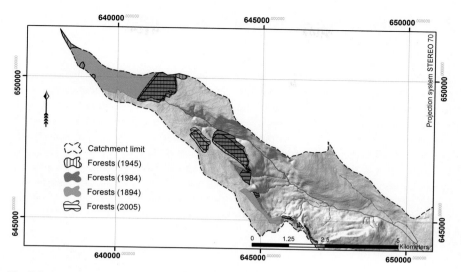

Fig. 7.2 Evolution of forested areas in the upper basin

wood was the main raw material used in construction (Băican 2001). For Iaşi county, the total area covered with forests occupies 73,504 ha (13.5 % of the total area of the county) (Lupaşcu and Onofrei 2009).

Referring strictly to the study area, at present, as in the past, the only wooded areas are found in the northwestern extremity part of the basin and occupies 1.8 % of the total area. The need for arable land, plant cultivation, initiating a defor- estation which resulted in triggering the degradation of soils: landslides, gully erosion and sheet erosion. It was observed (Fig. 7.2) and recorded a dramatic decrease in the early nineteenth century, from 425.75 to 182.07 ha, which can be associated with massive forest fires between the period from the two World Wars. After that period, the trend is the same, but not as in high proportions, from 182.07 ha in 1945 to 161.41 ha in 1984. However, in recent years amid public awareness of forest plantations being carried out by the Local Forestry Offices (like the one from Paşcani city), in the attempt to reduce the area of land affected by erosion, the surface occupied by the forest has grown, reaching 197.67 ha in 2006. It can be observed as an ascending trend of forest growth in recent years.

One of the few projects of afforestation from Iaşi county is in Cucuteni com- mune. As a result of a project with funding from the Ministry of Environment, the plateau area within the catchment will be afforested on an area of 53 ha. The investment is needed to halt and prevent the landslides that affect the village, as well as archaeological and palaeontological sites.

7.2 The Spread of Archaeological Sites According to Vegetation

The presence and automatisms of the forests have represented since ancient times places for human settlements. Besides the main exploitation role for its wood resources for the purpose of building and heating houses, contributing to the overall temperature regulation in the specific area (higher thermal comfort in the warm season and the role of shelter in the way of the winds in the cold season), also as shelter for wild animals hunted by the Neolithic populations. From the currently existing forest at the contact between the plain and plateau, specific species of the same family of trees have been preserved since the Chalcolithic: hornbeam (*Carpinus betulus*), elm (*Ulmus campestris*), ash (*Fraxinus excelsior*), wild cherry (*Cerasus avium*), within stands associations: corn (*Cornus mas*) and hazelnut tree (*Corylus avellana*).

References

Aniței LG (2000) Flora și vegetația Bazinului Bahlui (județul Iași). Ph.D. dissertation, Univ. "Al. I. Cuza" Iași
Băican V (2001) Spatial forest evolution in the Moldavian Plain after the cartography of the 18th and 20th centuries. An. șt. ale Univ. "Al. I. Cuza" XLVII(II):25–32
Ion C, Stoleriu CC, Baltag E, Mânzu C, Ursu A, Ignat AE (2011) Păsările și habitatele din zonele umede ale Moldovei. Editura Univ. "Al. I. Cuza" Iași
Lupașcu A, Onofrei M (2009) Characteristics of the forest fund in Iași county, with special reference to the forest wards of Iași and Ciurea. PESD 3:159–167
Minea I (2012) Bazinul hidrografic Bahlui. Studiu hidrologic. Editura Univ. "Al. I. Cuza" Iași

Chapter 8
Soils

Abstract The pedological complexity is determined by the specific lithologic complexity of the Moldavian Plain. In order to analyze the distribution of soil classes the pedological maps were digitized. Over the soil classes' layer the archaeological sites layer was overlapped to identify the spread of prehistoric settlements according to this factor; more than half of the settlements are placed on Chernisols, which represent soils with a high fertility. Crops held a highly important role in the development of the Chalcolithic civilization, agriculture being one of the main activities.

Keywords Chernozems · Paleosols · Pedogenesis · Sediment · SRTS

Soil is defined as the natural cover from the top of the lithosphere, in a continuous evolution, being under the action of pedogenetic factors (parent material, relief, groundwater, climate, vegetation and time) and anthropogenic activities. Each factor plays its role in the formation and evolution of soil types. Both formation and evolution of soil horizons and profiles, is done under the influence of several categories of processes: input or accumulation, output or loss, transformation, translocation, and mixing processes (Lupaşcu et al. 1998).

External conditions (relief and topography) directly affect soil properties such as moisture, coverage and variety of vegetation, altitude and aspect and the degree of infiltration of water into the soil (Bunting 1967). In an archaeological context, it is necessary to distinguish between the soil (which is static and formed in situ through various chemical and biological processes) and the sediments (which are dynamic, affected by erosion, transport and deposition in a certain environment).

8.1 Soil Classes and Types Distribution

Analysing the distribution of soil types and classes has been complied a spatial database using the pedological studies belonging to Bălţaţi, Belceşti, Cotnari, Cucuteni, Todireşt communes and Tg. Frumos city, scale 1:10,000, made by Iaşi

© The Author(s) 2016
I.C. Nicu, *Hydrogeomorphic Risk Analysis Affecting Chalcolithic Archaeological Sites from Valea Oii (Bahlui) Watershed, Northeastern Romania*,
SpringerBriefs in Earth System Sciences, DOI 10.1007/978-3-319-25709-9_8

Table 8.1 The distribution of
the main soil classes from
Valea Oii catchment

Class	S (ha)	% (from total)
Chernisols	6470.89	74.27
Luvisols	364.83	4.19
Hydrisols	124.71	1.43
Protisols	539.82	6.20
Antrisols	1052.72	12.08
Vertisols	159.81	1.83
Total	8712	100

County Office for Soil Survey in 1994, 1995, 1997 and 2003. They were previously scanned and georeferenced in STEREO 70 projection system. A database resulted of polygons drawn manually, having corresponding attributes table consisting of data structured in the class, type, subtype and specific properties of soils, being correlated with the areas generated automatically by tracing a vector layer. For the classification of soils SRTS (Romanian Soil Taxonomy System) (Florea and Munteanu 2012) was used. According to the final soil map of the catchment, by digitization has been obtained 400 polygons and 155 soil units.

It can be observed (Table 8.1) the clear domination of Chernisols which holds three quarters out of the basins surface 6470.89 ha (74.27 %). Second to follow are the soils formed under the anthropogenic influence—antrisols, with a surface of 1052 ha (12.08 %), protisols, which occupy about 539.82 ha (6.20 %) and also luvisols with 364.83 ha (4.19 %). In lower proportions, vertisols appear (159.81 ha, 1.83 %), with an azonal character, and hydrisols, occupy the smallest surface of 124.71 ha (1.43 %).

As soil type **chernozems** dominate, with an area of 5851.89 ha (67.16 %), with high fertility, encountered in the lowland plain area; they were formed due to climatic conditions, like temperate-continental with excessive influences typical for the eastern part of the country (Niacşu 2012), with annual average temperatures between 8 and 9.5 °C and rainfalls with values between 500 and 600 mm. The lithology is mostly constituted of loess and clays, scattered on plateaus and on gentle slopes on the left side of the basin.

Anthrosols, formed under anthropic influence, ranks second in the basin as spread, with an area 1052.72 ha (12.07 %); they are encountered especially on the right side of the basin, on the slopes affected by erosion processes, and where clays and loess deposits dominate as parent material. Also, they are found in the north-western part of Balş village and east from Bălţaţi village, where are vineyards and fruit trees (apples).

Phaeozems is the second type of soil representative within chernisol type, which were formed under the influence of herbaceous primary or secondary meso-hydrophite vegetation, that has maintained for a long time (Blaga 1996); occupies an area of 619 ha (7.1 %) encountered in the upper part of the basin, on Laiu plateau on surfaces originally occupied by forests and silty substrate.

Fluvisols are in the early stages of formation, encountered on a 380.51 ha surface (4.37 %) on the main course of the valley (in particular between Băiceni and Boureni villages and sometimes on some secondary valleys (Turcului Valley, Făcuți Valley, Babelor Valley). **Luvisoils**, with an area of 189.64 ha (2.18 %) are formed in close connection with forest vegetation, found exclusively in the upper part of the basin, in the plateau area. Then follows with lower percentage preluvisols (2.01 %) and regosols (1.59 %).

8.2 The Role of Soils in the Placement of Archaeological Sites

Knowledge of soils in archaeology is an essential factor due to the fact that soil represents the base on which the prehistoric man has evolved since yearly ages until the present, but also the element in which the soil traces are preserved. The soil must not be considered a special element, but it must be placed in a physical-geographical context. If in the analysis of the archaeological material found buried, a study of its sedimentary rock side, spatial repartition, modifications due to pedogenetic processes, then it can be considered that only a part of the archaeological information is being studied (French 2005), an insufficient factor for the realisation of a complete analysis. Before beginning archaeological digging, holding information regarding classes and types of soil, obtained through pollen analysis (Tipping et al. 1999), can facilitate the understanding of the placement of certain settlements in different places, from different historical periods.

Another research domain in which the soils proprieties have a significant value is the one of the geophysical prospection (magnetometry, soil electrical resistivity, ground penetrating radar). In the case of the GPR measurements (which functions on the principle of electromagnetic waves propagation in the soil), one of the most important properties is held by the electrical conductivity which depends directly on the high content of water, clay and soluble salts (McNeill 1980). The soils with a high content of salts generate an increase of electrical conductivity and thus are not suitable for GPR measurements (Doolittle and Collins 1995). In the case of these soils, the penetrations depth is constrained to less than 25 cm, while in normal conditions, the penetration depth (which depends on the utilisation frequency) can reach 25–30 cm (Doolittle and Butnor 2009). The same depth reduction marks were reported in the cases of soils with a high content of calcium carbonate ($CaCO_3$) (Grant and Schultz 1994), or the ones with a high content of clay particles because they have the propriety of retaining a higher quantity of water (particles with a dimension of <0.002 mm) (Olhoeft 1986).

Human communities have set their settlements there where they had noticed a better development of agricultural production, in places where the soil proprieties were able to be used and the existent resources which enabled them to survive. After the execution of the distribution map of soil classes, a vectorial layer with the

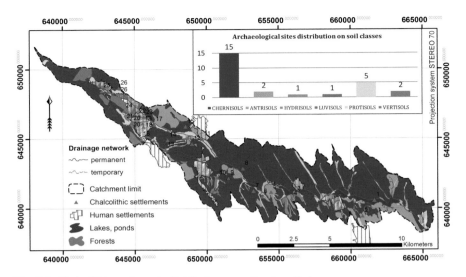

Fig. 8.1 Chalcolithic settlements distribution according to soil classes

Chalcolithic archaeological sites was overlapped. Thus, the placement analysis of archaeological sites according to the existence of soil classes was possible, our main interest being found in the fertile soils used for agriculture (Chernozems) and the soils with a high content in salts (a possible resource used on a larger scale by the prehistoric populations).

The fact that in the past (6000–2000 BC) there were existing the same classes and types of soils which could have influenced the placement of archaeological sites cannot be stated; however, major modifications did not take place. Exceptions are the areas covered in the past with forests, especially the upper part of the basin, in the plateau area, where mainly are found hardwood species: oak (*Quercus robur*), elm (*Ulmus lamellosa*) and hornbeam (*Carpinus betulus*).

From Fig. 8.1 it can be observed that a number of 15 archaeological sites are presently placed on Chernisols (soils with a high fertility) which even if in the past were in the forming stage were construed for developing mixed vegetation associations, crops held a highly important role in the development of the Chalcolithic civilization, agriculture being one of the main activities (Boghian 2004). The sites from this category are distributed almost uniformly on the entire surface of the basin, with a plus for the superior basin, where most of the settlements are concentrated. In the case of the Protisols class, 4 sites are placed in this class, with soils still in an incipient stage of formation, mainly in meadows. The sites which are found on Antrisols (2 sites) are mainly affected by the slope processes, being under strong anthropic influence. Thus results the importance of internal soil properties, because the degree of vegetation coverage partly influences the erosion processes with negative effects among actual sites degradation. The effects are visible, when significant quantities of archaeological material are being washed and brought at the base of the slopes by the rain waters.

The soil is in continuous formation and evolution process, being difficult for us to estimate, and estimations at a stage of assumptions, the extension and presence of a certain type of soil for the prehistoric periods. In the analysis of this element, the close study of internal factors (rock type), external (climate, hydrological regime) and anthropic factors, which condition the processes of deposition or erosion, can help us in estimating or redoing the soil layer from the past.

Along time, stratigraphic studies held in different archaeological sites in which systematic diggings took place have brought a significant contribution to the chronology and evolution of the culture. A well-preserved and conserved soil (where agricultural works did not take place in an intensive manner, a rare aspect though) can offer crucial information regarding the stored archaeological material. The stratigraphic analysis is one of the ground methods of archaeological research, which especially indicates succession and not duration.

References

Blaga G, Rusu I, Udrescu S, Vasile D (1996) Pedologie. Editura Didactică și Pedagogică, R.A., București

Boghian D (2004) Comunitățile cucuteniene din bazinul Bahluiului. Editura Universității "Ștefan cel Mare" Suceava

Bunting BT (1967) The geography of soils. Hutchinson, London

Doolittle JA, Collins ME (1995) Use of soil information to determine application of ground-penetrating radar. J Appl Geophys 33:101–108

Doolittle JA, Butnor JR (2009) Soils, peatlands, and biomonitoring. In: Jol HM (ed) Ground penetrating radar. Theory and applications. Elsevier

Florea N, Munteanu I (2012) Sistemul Român de Taxonomie a Solurilor (SRTS). Editura SITECH, Craiova

French C (2005) Geoarchaeology in action: studies in soil micromorphology and landscape evolution. Routledge

Grant JA, Schultz PH (1994) Erosion of ejecta at Meteor Crater: constraints from ground-penetrating radar. In: Proceedings of fifth international conference on ground-penetrating radar, June 12–14, 1994, Kitchner, Ontario, Canada, Waterloo Centre for Groundwater Research and the Canadian Geotechnical Society, pp 789–803

Lupaşcu G, Jigău G, Vârlan M (1998) Pedologie generală. Editura Junimea, Iași

McNeill JD (1980) Electrical conductivity of soils and rock. Technical Note TN-5:21, Geonics Limited, Mississauga, Ontario, Canada

Niacşu L (2012) Bazinul Pereschivului (Colinele Tutovei). Studiu de geomorfologie și pedogeografie cu privire specială asupra utilizării terenurilor. Editura Univ. "Al.I.Cuza" Iași

Olhoeft GR (1986) Electrical properties from $10 - 3$ to $10 + 9$ Hz—physics and chemistry. Physics and chemistry of porous media II. In: Proceedings of american institute of physics conference, Ridgefield, Connecticut, USA, pp 281–198

Tipping R, Long D, Carter S, Davidson D, Tyler A, Boag B (1999) Testing the potential of soil-stratigraphic palynology in podsols. In: Pollard AM (ed) Geoarchaeology: exploration, environments, resources. Geological Society, London, Special Publication No. 165

Chapter 9
Geoarchaheology or Archaeogeomorphology?—Border Sciences

Abstract In this chapter a short history and evolution of geoarchaeology is provided. Being an interdisciplinary science, there are difficulties in finding and providing a precise definition, even for the experienced researchers. Worldwide there is a strong tradition in geoarchaeological studies published in dedicated journals and a lot of active working groups within professional societies (WGG, as a part of the IAG). That is why is easily confused with archaegeomorphology, which has a shorter history, but a high development potential. In the end, this study is identified as being an archaegeomorphological one, rather than a geoarchaeological one.

Keywords Archaeogeomorphology · Geoarchaeology · Human–environment relation · Interdisciplinarity · Landscape evolution

Geoarchaeology, through the mediation of techniques and methods of research belonging to earth sciences, has specifically focused on in situ studies of the deposition conditions of sediments and the formation processes. The study of archaeological sites in a context in which the conditions of the environment are being studied and analysed (Wandsnider 1992), or the ones which include punctual archaeological research, requires a geoarchaeological analysis at a regional scale, that must commence at the beginning of the research. Therefore, any old document or records related to the natural environment conditions, need to be studied and integrated among the archaeological research. This allows for a more efficient identification of possible connections (if there are any) between the two fields of study.

The human–environment relation is the one closely tied and interdependent, since humans, or communities of humans, have always taken in consideration, with or without their will, the characteristics of the environment (*geological conditions* —basement resources of raw material for raising domestic settlements, production of hunting weapons, using flint as raw material, places of salt resources exploitation; *geomorphologic conditions*—the placement of settlements on structural plateaus in a defensive purpose, or the defence towards natural hydrological phenomena like floods, in contact areas for facilitating mobility between certain communities, exposition towards the Sun; *hydrological conditions*—proximity

© The Author(s) 2016
I.C. Nicu, *Hydrogeomorphic Risk Analysis Affecting Chalcolithic Archaeological Sites from Valea Oii (Bahlui) Watershed, Northeastern Romania*,
SpringerBriefs in Earth System Sciences, DOI 10.1007/978-3-319-25709-9_9

towards water supply resources—ponds, salty ponds, water streams; *pedological conditions*—soil fertility, mineral resources, the existence of consistent clay resources used in pottery; *vegetation and fauna conditions*—the existence of a rich forest fund which also constitutes a place of existence for wild animals, used as raw material for building houses, heating them in the cold season, food preparation, but also burning pottery) (Nicu et al. 2012).

The achievement of these connections has evolved during time and has become imperative in the study of archaeology (Table 9.1), which encompasses environment description, archaeological monographs until the studies related to the landscape archaeology (Aston 2002), environmental archaeology, ethno-archaeology (David and Kramer 2001) and geoarchaeology (Wilson 2011).

On the other hand, the impact humans have on the environment has become increasingly problematic, from an archaeological point of view, but equally from a sustainable development perspective (Goldberg and Macphail 2006), which includes challenges such as climate change, deforestation, desertification, soil erosion etc. Here, geoarchaeology steps in, which studies the traces and impact of human interaction with the environment since early times until present days. Geoarchaeological research contributes in two equally important ways, it supplies data concerning changes that took place in a certain region and it permits the reconstruction of old landscapes and understanding paleoclimatic evolution. The evolution of humans has involved the exploitation of resources provided by the geosphere, which has produced changes, sometimes irreversible to the environment.

The discovery of new archaeological sites must be managed sustainably to ensure their long-term protection and to enable them to be available for future generations. Humans are capable of surviving provided by their capacity to adapt. This is one of the basic abilities of humans, whether they are prehistoric or modern. Another conception of geoarchaeology is the identification and investigation of at least three main topics:

Table 9.1 Basic components of geoarchaeology (Brown 2001)

Component	Specific methods of research
1. Locating archaeological sites	Topographic maps, remote sensing, GIS analysis
2. Geomorphologic analysis of relief	Mapping, stratification, dating
3. Stratigraphic studies	Combining of geomorphologic studies with remote sensing
4. Analysis of sedimentary deposits	Identification of the geological substrate (mineral studies, texture analysis, etc.)
5. Paleogeographic analysis	Analysis of geological substrate, paleoecology (snails, pollen, wood, insects, seeds)
6. The determination of the human–environment relations	Connections between environment and the cultural changes, cost surface analysis, catchments analysis
7. Natural hazards studies	The majority of the above mentioned
8. Dating	Luminescence dating, C^{14} dating paleomagnetic dating

First   recognition and decoding of landforms formation, evolution and transformation. Within this framework, the effect of tectonic movements, sea
level changes (where applicable) and the effect that had on the actual relief
can be studied. In what ways was the actual relief affected in the past and if
it is possible to identify paleo-sequences in the present context? Are we
able to find possible connections between these events or processes?
Finally, connecting, relating and the issue of certain assumptions about the
evolution of the environment or landscape.

Second   is it possible to identify the effects that human activities have had on the
formation and evolution of the landscape? If there were links between the
characteristics of climate, landforms, soils and humans. Finally, the aim
being the attempt to reconstruct the evolution of the landscape, the categories of land use and identifying the relationship between climate, soils
and humans.

Third   which is the effect that hydrological regime has had on the landscape
evolution, the sediment load and the effect on archaeological sites (French
2005).

Geoarchaeology, being an interdisciplinary field of study, overcomes the
boundaries imposed by a single science, by combining interdisciplinary research
methods, new theoretical and practical approaches and perspectives.

In recent years, a new concept was born—*archaeogeomorphology*; the term was
used for the first time by the American archaeologist, Wandsnider (1992), but
without a continuation of the research and without using the term for the combination of the two fields of study (history and geography). The analysis of this term
is continued by Thornbush (2012), who supports the classification of archaeogeomorphology as a subdiscipline of applied geomorphology, the emergence of a new
publication dedicated to archaeogeomorphology, and recognising the contributions
made by geomorphologists in archaeology.

Far from being a work that deals with terminology issues, after analysing both
terms and the state "where excavation is lacking and there is no real collaboration
with archaeologists, only consultation, this work should be considered to be
essentially archaeogeomorphological" (Thornbush 2012), it can be stated that this
study is more likely to be an archaeogeomorphological one, rather than a geoarchaeological one. As stated above, these issues are analysed in detail with pros and
cons by Thornbush (2012).

9.1   Definitions

In the British accepted view, <u>geoarchaeology</u> represents a new field of research that
has met a fast development in the last decade. Actually, geologists were applying
methods and principles of archaeological research since 1863, then being recorded
the first links between earth sciences and archaeology (Lyell 1863).

In *Elsevier's Dictionary of Geography*, the term <u>*geoarchaeology*</u> is defined as a combination of cultural, economic, geological, paleo-geographical analysis for having the ability to determine the relations between human society from the past and the environment (Kotlyakov and Komarova 2007). An incomplete definition we can state, since the analysis and involved sciences are of a greater number.

Shackley (1979) established one of the first markers and attempts of defining <u>*geoarchaeology*</u>, as representing the application of earth sciences, including geo-physical prospections and petrographical analysis, the majority of research from this field having connections with geology, geomorphology, pedology, sedimentology.

A concise definition is that <u>*geoarchaeology*</u> represents a multiple relations approach, where methods and concepts from geography and geosciences find their applicability for studying prehistory, archaeology and history (Rapp and Hill 1998).

French (2005) adopts a scientific approach in defining the term, dedicating an entire volume of methods, problems and scopes, in order to offer a clear definition. In defining the term, he refers firstly to geomorphology (which represents the science that studies relief, genesis, evolution, dynamics, reports with human society). *Geoarchaeolgy* represents the combined study between archaeology's characteristics and geomor-phology and the "fingerprints" of the anthropic actions upon landscape evolution.

In defining the term <u>*geoarchaeology*</u>, an important contribution is held by Brückner, with a well-elaborated definition. It is one of the most complete in the acceptance that <u>*geoarchaeology*</u> is an interdisciplinary science by excellence, which combines the geo-bio-archives study in an archaeologic context with the help of geosciences for reconstructing the evolution and the utilisation of landscapes and ecosystems. The definition has special regard to the interaction between people and the environment, with objectives, perspectives and methods of natural sciences: geosciences (geology, sedimentology, mineralogy, etc.), physical geography (geomorphology, pedology, geoecology, biogeography, hydrology, meteorology and climate changes) and human sciences: archaeology, classic archaeology, historic sciences, prehistory, oriental studies, human geography (urban geography, rural geography, settlement geography, historical geography etc.) (Brückner and Vött 2008).

Research methods are complex, ranging from the classic (subsurface survey, excavations), up to the modern (non-invasive methods, where it can be included magnetometry, GPR, fluxgate, soil resistivity), all currently embedded in GIS (Wescott and Brandon 2005). All this for a better record and unification of data collected, for the exchange of information between the staff carrying non-invasive prospections and archaeologists carrying out the excavations to be easier.

The Working Group on Geoarchaeology, defines <u>*geoarchaeology*</u> as "the geo-sciences and geographical methods and techniques applied to prehistory, archae-ology, and history" (Fouache et al. 2010).

All data obtained (with the correspondent in the real environment) are integrated into the database in the form of maps, plans and profiles, with the help of GIS (Harrower 2010; Mehrer and Wescott 2006), which is also used in the discovery of new archaeological sites, through archaeological predictive modelling (Verhagen 2007; Graves 2011; Balla et al. 2013; Stirn 2014) or to confirm those already discovered, through satellite images (Goossens et al. 2006).

9.2 Evolution

Using classic archaeology as a starting point, there were researchers who made different connections between the existence and the placement of archaeological sites and the factors which determined their placement. The link between archaeology and the environment (geological factor, climatic factor) took place since 1863 (Lyell), the precursors which led to the development and implementation of new research methods were: Brakenridge (1986), Bryson (1994), Hubert (2001), Kirkby and Kirkby (1976), McGlade (1995).

A fast development took place in the paleo-geomorphological reconstruction, both on the basis of numerical models of terrain and modern methods, of non-destructive archaeological prospections (Ground Penetrating Radar, magnetometer, soil resistivity), but also from carbon dating: Stafford (1995), Brückner (2003), Ghilardi et al. (2008). Combining and utilising these methods from physics, chemistry, biology and geography are called *archaeometry*, which is a term introduced in the literature by Prof. Christopher F.C. Hawkes from the Oxford University, England, in the year 1985.

Today, geoarchaeology is a self-standing study domain, with results being reported in books and journals, which had their debut in the 1950s–1960s with titles such as: *Quaternary Research* (1963), *Journal of Archaeological Science* (1973), *Archaeomaterials* (1985), *Geoarchaeology—An International Journal* (1986), *Archaeological Prospection* (1995), *S.A.P.I.EN.S*, but also in publications which have as main subjects archaeology, anthropology or geology, such as: *Journal of Human Evolution, Journal of Sedimentary Research, American Antiquity, Antiquity*, etc. Specialised publications are dedicated to methodology, used in the study of geoarchaeology: *X-Ray Fluorescence Spectrometry (XRF) in Geoarchaeology, Journal of Archaeological Method and Theory, Archaeological Prospection, Remote Sensing in Archaeology, GIS and Archaeological Site Location Modelling*, etc. International symposiums and congresses dedicated to geoarchaeology were also popularised: ISA (*International Symposium of Archaeometry*), *First, Second and Third Arheoinvest Congress—Interdisciplinary Research in Archaeology* etc.).

In the evolution and development, a significant contribution is held by the establishment within the IAG (International Association of Geomorphologists) of the WGG (Working Group on Geoarchaeology), in 1997; the groups activity consisted of organising conferences dedicated to this field: *Geoarchaeology of the Landscapes of Classical Antiquity* (1998, Belgium), *Geoarchaeology in Northwestern Europe* (1999, UK), *the International Colloquium on Geoarchaeology. Landscape Archaeology. Egypt and the Mediterranean World* (2010, Egipt), *Geomorphic processes and geoarchaeology. From Landscape Archaeology to Archaeotourism* (2012, Russia); or dedicated sessions within international congresses: Geoarchaeology within global environmental change (2014, EGU, Austria), *Geoarchaeology: Human–environment interactions in the Pleistocene and Holocene* (2015, EGU, Austria), *Geoarchaeology: Human–*

environment interactions and palaeo-geohazards (2015, INQUA, Japan). As it can be observed, the groups' preoccupations are diverse and have as a scope the development and implementation not only of research methods, but also publishing the results (source: http://www.geomorph.org/wg/wgga.html. Accessed on 21 May 2015).

In contrast to the stage of foreign results and research, Romania's contribution has been lacking enthusiasm, summarising only some gradual research without an international echo. This is due, in part, to the lack of interdisciplinary funding, and the lack of qualified personnel. In 1996, during an initiative on the behalf of a researchers group from the Romanian National History Museum, the Pluridisciplinary National Research Center was established (Popovici et al. 2002).

Of course, the city of Iasi was not indifferent to geoarchaeological and archaeogeomorphological research, so well represented through some well-known historians; thus, the interdisciplinary research have started since 1987, when at the initiative of Prof. Mircea Petrescu-Dîmbovita, the *Cercetări interdisciplinare în arheologie* section appeared in the *Moldavian Archaeology* periodical. Later on the *Centrul Interdisciplinar de Studii Arheoistorice* (CISA), in the "Al. I. Cuza" University of Iași, the center being established with the aim of "establishing contacts and collaborations with all that wish to contribute to the progress of the interdisciplinary archaeological research" (Ursulescu 2006). All this culminated with the establishment, during the *Platformei de formare și cercetare în domeniul arheologiei, Arheoinvest Platform*, of the Geoarchaeology laboratory, after winning a research grant; we can state, because I am a member in the Platform, that it holds the necessary logistic support, including qualified personnel, necessary in optimal ongoing conditions and international standards of research.

Further, we can add the Pluridisciplinary National Research Center ("Valahia" University of Târgoviște), *Computerized Archaeology Department* (National History of Transilvania Museum from Cluj-Napoca), and *Institutul de Cercetări Eco-Muzeale* (Tulcea). The development of these centers, technologic progress, summer schools in this field and auxiliary qualified personnel have built the premises of collaborations between Romanian and foreign researchers (Maillol et al. 2004; Micle et al. 2010; Carozza et al. 2012, 2014; Nicu et al. 2012; Iovita et al. 2013).

Attention was particularly directed to identify human–environment relation, identifying natural resources from a specific territory, the relief patterns and the role of the morphometric and geomorphologic factor in placing the settlements and paleo-geographic reconstructions (Văleanu 2003; Boghian 2004; Bounegru et al. 2009; Micle 2011; Măruia 2011; Asăndulesei 2012; Romanescu 2013; Sherwood et al. 2013; Romanescu et al. 2015), the role of natural hazards in prehistoric populations dynamics (Nicu and Romanescu 2015).

The people perception regrading the production and development of natural phenomenons has had a sinuous evolution along time. Usually, when man intervenes in the path of nature, it has a tendency of coming back to the initial path, sometimes through some phenomenas perceived by man as dangerous. At the beginnings, these phenomenas were considered as "acts made by God" to punish mankind for their sins, but this conception was overcome with solid arguments by Immanuel Kant and Jean-Jacques Rousseau; their vision towards these events was a realistic one, and looking at disasters as natural events, mankind was responsible since it intervened with nature attempting to change it (Huggett 1997).

United Nations Educational, Scientific and Cultural Organization (UNESCO, 1946) and the Council of Europe represent the key global players concerned with the protection of the tangible heritage across the world, and the assessment of the natural and anthropic risks directly affecting it. In their turn, these organisations founded dedicated institutions for preserving the cultural and natural heritage, of which: International Council for Monuments and Sites (ICO-MOS) is the most prominent, followed by International Committee for Architectural Photogrammetry (CIPA), International Society for Photogrammetry and Remote Sensing (ISPRS), etc. Their attention turns, foremost, to restoring and preserving the sites or monuments already in an advanced state of degradation, and seldom with prevention (Catani et al. 2002; Georgopoulos and Ioannidis 2004).

The protection and preservation of the archaeological heritage in Romania is a sensible issue, both on account of the lack of a complete registry of existing sites, but also due to the continuous degradation caused by active processes of natural erosion and anthropic activity. Notable work has been dedicated during the last years to identifying and adding as many sites as possible into the National Archaeological Registry (RAN) managed directly by the Ministry of Culture, and into the official database of archaeological heritage developed by the Institute of Cultural Memory (CIMEC) and the National Heritage Institute (INP). Even though between years 2000 and 2013 RAN lists 17,700 sites and 30,000 archaeological entities from 5626 localities (Oberländer-Târnoveanu 2014), Romania is still far from having a complete database of the archaeological heritage. From the total, Iași county (with a surface area of 5476 km^2), where the study area is located, only has 245 sites listed, compared with Sălaj county (with an area of 3864 km^2), which at 746 sites has the largest number of entries in the country (www.cimec.ro. Accessed on 25 May 2015).

References

Asăndulesei A (2012) Aplicații ale metodelor geografice și geofizice în cercetarea interdisciplinară a așezărilor cucuteniene din Moldova. Studii de caz. PhD Dissertation, Univ. "Al. I. Cuza" Iași

Aston M (2002) Interpreting the landscape. Landscape archaeology and local history. Routledge, London

Boghian D (2004) Comunitățile cucuteniene din bazinul Bahluiului. Editura Universității "Ștefan cel Mare" Suceava

Balla A, Pavlogeorgatos G, Tsiafakis D, Pavlidis G (2013) Locating Macedonian tombs using predictive modelling. J Cult Herit 14(5):403–410. doi:10.1016/j.culher.2012.10.011

Bounegru O, Romanescu G, Alexianu M, Dumitrache I, Vasiliniuc I (2009) Hinterland and site catchment studies at Histria on the Black Sea coast, Romania. Antiquity 83(322), http://www.antiquity.ac.uk/projgall/bounegru322/

Brakenridge GR, Schuster J (1986) Late quaternary geology and geomorphology in relation to archaeological site locations, Southern Arizona. J Arid Environ 10:225–239

Brown AG (2001) Alluvial geoarchaeology. Floodplain archaeology and environmental change. Cambridge University Press, Cambridge

Brückner H (2003) Uruk—a geographic and paleo-ecologic perspective on a famous ancient city in Mesopotamia. Bensheim, Geooko, Band/vol XXIV, pp 229–248

Brückner H, Vött A (2008) Geoarchäologie – eine interdisziplinäre Wissenschaft par excellence.– In: Kulke E, Popp H (eds) Umgang mit RisikenKatastrophen – Destabilisierung – Sicherheit. Tagungsband Deutscher Geographentag 2007 Bayreuth. Herausgegeben im Auftrag der Deutschen Gesellschaft für Geographie. S. 181–202, Bayreuth, Berlin

Bryson RA (1994) On integrating climatic change and culture change studies. Hum Ecol 22 (1):115–128. doi:10.1007/BF02168766

Carozza JM, Micu C, Mihail F, Carozza L (2012) Landscape change and archaeological settlements in the lower Danube valley and delta from early Neolithic to Chalcolithic time: a review. Quatern Int 261:21–31. doi:10.1016/j.quaint.2010.07.017

Carozza JM, Carozza L, Radu V, Leveque F, Micu C, Burens A, Opreanu C, Haita C, Danu M (2014) After the flood: geomorphological evolution of the Danube delta after the Black sea— Mediterranea reconnection and its implications on Eneolithic/Chalcolitique settlements. Quaternaire 24(4):503–512

Catani F, Fanti R, Moretti S (2002) Geomorphologic risk assessment for cultural heritage conservation. In: Allison RJ (ed) Applied geomorphology. Wiley, Chichester

David N, Kramer C (2001) Etnoarchaeology in action. Cambridge University Press, Cambridge

Fouache É, Pavopoulos K, Fanning P (2010) Geomorphology and geoarchaeology: cross-contribution. Geodin Acta 23:207–208. doi:10.1080/09853111.2010.9736394

French C (2005) Geoarchaeology in action: studies in soil micromorphology and landscape evolution. Routledge, London, pp 3–10

Georgopoulos A, Ioannidis C (2004) Photogrammetric and surveying methods for the geometric recording of archaeological monuments. In: Archaeological surveys, FIG working week, Athens, Greece

Ghilardi M, Fouache E, Queyrel F, Syridres G, Vouvalidis K, Kunesch S, Styllas M, Stiros S (2008) Human occupation and geomorphological evolution of the Thessaloniki Plain (Greece) since Mid Holocene. J Archaeol Sci 35(1):111–125. doi:10.1016/j.jas.2007.02.017

Goldberg P, Macphail RI (2006) Practical and theoretical geoarchaeology. Blackwell Science Ltd

Goossens R, De Wulf A, Bourgeois J, Gheyle W, Willems T (2006) Satellite imagery and archaeology: the example of CORONA in the Altai Mountains. J Archaeol Sci 33:745–755. doi:10.1016/j.jas.2005.10.010

Graves D (2011) The use of predictive modelling to target Neolithic settlement and occupation activity in mainland Scotland. J Archaeol Sci 38(3):633–656. doi:10.1016/j.jas.2010.10.016

Harrower MJ (2010) Geographic information system (GIS) hydrological modeling in archaeology: an example from the origins of irrigation in Southwest Arabia (Yemen). J Archaeol Sci. doi:10.1016/j.jas.2010.01.004

Hubert S (2001) Modélisation Numérique de Terrain et analyse spatiale pour une aide à l'évaluation des risques d'inondation dans la region de Nouakchott (Mauritanie). Mémoire de DESS, ENSG – Univ. Pierre et Marie Curie, 50

Huggett R (1997) Catastrophism: asteroids, comets and other dynamic events in earth history. In: Bryant E (ed) Natural hazards. Cambridge University Press, Cambridge

Iovita R, Dobos A, Fitzsimmons KE, Probst M, Hambach U, Robu M, Vlaicu M, Petculescu A (2013) Geoarchaeological prospection in the loess steppe: preliminary results from the Lower Danube Survey for Paleolithic Sites (LoDanS). Quatern Int 351:98–114. doi:10.1016/j.quaint.2013.05.018

Kirkby A, Kirkby MJ (1976) Geomorphic processes and the surface survey of archaeological sites in semi-arid areas. In: Davidson DA, Shackley ML (eds) Geoarchaeology: earth science and the past. Westview Press, Denver

Kotlyakov VM, Komarova AI (2007) Elsevier's dictionary of geography

Lyell C (1863) Geological evidences of the antiquity of man. John Murray, London

Maillol JM, Ciubotaru DL, Moravetz I (2004) Electrical and magnetic response of archaeological features at the early neolithic site of movila lui Deciov, Western Romania. Archaeol Prospect 11:213–226. doi:10.1002/arp.234

Măruia IL (2011) Cercetări interdisciplinare vizând reconstituirea geografiei istorice a Dealurilor Lipovei. Excelsior Art

McGlade J (1995) Archaeology and the ecodynamics of human-modified landscapes. Antiquity 69:113–132

Mehrer MW, Wescott KL (eds) (2006) GIS and archaeological site location modeling. CRC Press, Taylor and Francis Group

Micle D, Măruia L, Török-Oance M, Lazarovici G, Mantu Cornelia-Magda, Cîntar A (2010) Archaeological geomorphometry and geomorphography. Case study on Cucuteni—a site from Ruginoasta and Scânteia, Iaşi County, Romania. Annales d'Université Valahia Târgovişte, Section d'Archéologie et d'Histoire, XII(2):23–37, Târgovişte

Micle D (2011) Un model practic de aplicare a topografiei şi cartografiei arheologice în analiza spaţială a habitatului post-roman din Dacia de sud-vest între sfârşitul secolului al II-lea şi începutul secolului al V-lea p. Chr. Excelsior Art, Timişoara

Nicu IC, Asăndulesei A, Brigand R, Cotiugă V, Romanescu G, Boghian D (2012) Integrating geographical and archaeological data in the Romanian Chalcolithic. Case study: Cucuteni settlements from Valea Oii (Sheep Valley—Bahlui) watershed. Geomorphic processes and geoarchaeology. From landscape archaeology to archaeotourism, Moscova—Smolensk, <<Universum>>:204–207

Nicu IC, Romanescu G (2015) Effect of natural risk factors upon the evolution of Chalcolithic human settlements in Northeastern Romania (Valea Oii watershed). From ancient time dynamics to present day degradation. Z Geomorphol. doi:http://dx.doi.org/10.1127/zfg/2015/0174 (in press)

Oberländer-Târnoveanu I (2014) Patrimoniul arheologic naţional: politici, documentare, acces. In: Musteaţă S (ed) Arheologia şi politicile de protejare a patrimoniului cultural în România – culegere de studii, Chişinău-Iaşi

Popovici D, Bălăşescu A, Haită C, Radu V, Tomescu M, Tomescu I (2002) Cercetarea arheologică pluridisciplinară. Concepte, metode şi tehnici. Ed. Cetatea de Scaun, Târgovişte

JrG Rapp, Hill CL (1998) Geoarchaeology: the earth-science approach to archaeological interpretation. Yale University Press, New Haven

Romanescu G (2013) Geoarchaeology of the ancient and medieval Danube Delta: Modeling environmental and historical changes. A review. Quat Int 293:231–244. doi:10.1016/j.quaint.2012.07.008

Romanescu G, Bounegru O, Stoleriu CC, Mihu-Pintilie A, Nicu IC, Enea A, Stan CO (2015) The ancient legendary island of PEUCE—myth or reality? J Archaeol Sci 53:521–535. doi:10.1002/gea.21434

Sherwood SC, Windingstad JD, Barker AW, O'Shea JM, Sherwood WC (2013) Evidence of Holocene Aeolian activity at the close of the middle bronze age in the Eastern Carpathian Basin: geoarchaeological results from the Mureş River Valley, Romania. Geoarchaeol Int J 28 (2):131–146. doi:10.1002/gea.21434

Shackley ML (1979) Geoarchaeology: polemic on a progressive relationship. Naturwissenschaften 66(9):429–432

Stafford CR (1995) Geoarchaeological perspectives on paleolandscapes and regional subsurface archaeology. J Archaeol Method Theory 2(1):69–104. doi:10.1007/BF02228435

Stirn M (2014) Modeling site location patterns amongst late-prehistoric villages in the Wind River Range, Wyoming. J Archaeol Sci 41:523–532. doi:10.1016/j.apgeog.2011.12.005

Thornbush MJ (2012) Archaeogeomorphology as an application in physical geography. Appl Geogr 34:325–330. doi:10.1016/j.apgeog.2011.12.005

Ursulescu N (2006) Cercetarea arheologică interdisciplinară în centrul universitar Iaşi şi unele probleme actuale şi de perspectivă ale arheologiei. In: Popovici D, Anghelinu M (eds) Cercetarea arheologică pluridisciplinară în România. Trecut, prezent, perspective, Cetatea de Scaun, Târgoviște

Văleanu MC (2003) Omul și mediul natural în neo-eneoliticul din Moldova. Editura Helios, Iași

Verhagen P (2007) Case studies in archaeological predictive modelling. Leiden University Press

Wandsnider L (1992) Archaeological landscape studies. In: Rossignol J, Wandsnider L (eds) Space, time, and archaeological landscapes. Plenum Press, London

Wilson L (ed) (2011) Human interactions with the geosphere: the geoarchaeological perspective. Geological Society, London (Special Publications)

Wescott KL, Brandon RJ (eds) (2005) Practical applications of GIS for archaeologists. A predictive modelling toolkit. Taylor & Francis, London

Chapter 10
Archaeological Inventory

Abstract A short and concise analyze of the twenty six Chalcolithic archaeological sites is made in this chapter. Along the four stages of evolution (Precucuteni, Cucuteni A, Cucuteni A-B, Cucuteni B), the Chalcolithic civilization has spread over a large surface in Eastern Europe (approximately 350 000 km^2), being characterised by the development of brass and the assertion that ceramics were now starting to be decoratively painted. In order to understand the dynamics of the population along the four stages of evolution, the theory of Island Biogeography is used as a novelty.

Keywords Eneolithic · Evolution · Linear pottery · Migration · Pruth

Eneolithic (lat. *aenus* = copper) or *Chalcolithic* (gr. *chalkos* = brass, *lithos* = stone), after the Neolithic era, between 6000 and 2000 BC, is characterised by the development of brass and the assertion that ceramics were now starting to be decoratively painted (Ursulescu 1999).

The cultural complex Cucuteni-Ariușd-Trypillia (name given after the eponymous resorts from Ariușd—close by Sf. Gheorghe, Cucuteni—close to Târgu Frumos and Trypillia—in Ukraine, in close proximity to Kiev) spread along a large surface, of approximately 350,000 km^2 (on the Romanian, Moldavian and Ukrainian territories), from south-east of Transylvania (Brașov and Ciuc depression), Moldavian Sub-Carpathians floodplains, with entrances through the Oriental Carpathians, Curvature Sub-Carpathians and Moldavian Plateau valleys, until the plain areas between the interfluves Pruth-Dniestr, Dniestr-Bug, South Bug and Dniepr (Fig. 10.1). High elevation (hills, foothills, platforms, snouts hill, and terraces) was a key component in determining the placement of settlements (Monah 1985; Petrescu-Dîmbovița 2001).

At the international level, the Neolithic period research through the interdisciplinary methods are becoming a vast sought-after topic. On the basis of multilayer research in the settlement from Cucuteni, the first periodic scales were performed by Schmidt (Petrescu-Dîmbovița 1966).

The most recent absolute chronology data was obtained through palynology analysis and carbon dating throughout some representative settlements for the

© The Author(s) 2016
I.C. Nicu, *Hydrogeomorphic Risk Analysis Affecting Chalcolithic Archaeological Sites from Valea Oii (Bahlui) Watershed, Northeastern Romania*,
SpringerBriefs in Earth System Sciences, DOI 10.1007/978-3-319-25709-9_10

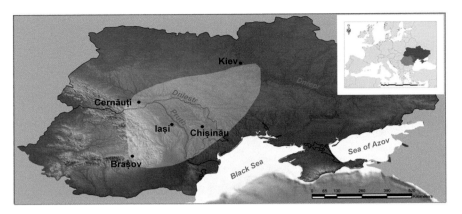

Fig. 10.1 The spreading area of Cucuteni-Ariușd-Trypillia cultural complex

Chalcolithic period (Cucuteni—*Cetățuie*, Iași county, Poduri—*Dealul Ghindaru*, Bacău county, Hăbășești—*Holm*, Iași county); the chronologic dating was and still is a real challenge for the archaeologists. The Cucuteni culture phases are (Mantu 1995, 1998):

- Precucuteni I: ~5050–4950 BC
- Precucuteni II: ~4950–4750 BC
- Precucuteni III: ~4750–4600/4550 BC
- Cucuteni A: ~4600/4550–4050 BC

  - Cucuteni A_1: ~4600–4550 BC
  - Cucuteni A_2: ~4550–4300 BC
  - Cucuteni A_3: ~4300–4150 BC
  - Cucuteni A_4: ~4150–4050 BC

- Cucuteni A–B: ~4050–3775 BC
- Cucuteni B: ~3775–3500 BC

  C_{14} dating indicated the below chronology, slightly different from the one above:

- Cucuteni A: ~4525/4500–3950 CAL BC

  - Cucuteni A_1: ~4525/4500–4450 CAL BC
  - Cucuteni A_2: ~4450–4150 CAL BC
  - Cucuteni A_3: ~4450–3800 CAL BC
  - Cucuteni A_4: ~4250–3950 CAL BC

- Cucuteni A–B: ~4050–3700 CAL BC
- Cucuteni B: ~3800/3750–3500/3450 CAL BC (Bem 2000).

10.1 Precucuteni Culture

As the name shows, this culture is leading up to that of the painted pottery, which is part of the great complex Cucuteni-Ariusd-Tripolie. The name was given by Professor Radu Vulpe in 1936 during excavations from Izvoare (Neamt county) (Dumitrescu 1957). Early Chalcolithic culture was represented by a synthesis between central-eastern Europe, where tribes were known for their linear pottery and the culture of Boian—Giulesti communities (Petrescu-Dîmboviţa 2001).

Representative for phase **I**, the advanced manufacturing of pottery, in the settlement from Traian—*Dealul Viei*; was widely spread in western and central Moldavia as well as south-eastern Transylvania (Petrescu-Dîmboviţa 2001), the pottery is characterised by an incised decoration, usually spiralled, with narrowed lines, marked by two parallel, vertical lines, or fine points, round, triangular or stitches of points larger or smaller, irregularly shaped or oval (Marinescu-Bîlcu 1974).

In phase **II** took place the almost permanent sketching of Precucuteni culture, stretching across the northern half of Moldavia, and to the East, almost to the river Dnestr (Ursulescu 1999). Specifically their good quality well-burnt clay which made high-resistent pottery, in different tones of grey, brown-grey, usually black-grey or black inside the vessels (Marinescu-Bîlcu 1974).

During phase **III**, communities started to extend their area of settlement towards Dnepr, including the forest steppe area east of Dnestr river, having the name, in the territory of the present day Ukraine–*Tripolie A*. To the south, Precucuteni communities have been in contact and had ties with Hamangia and Boian culture; from this combination between the cultural aspects from north-eastern Muntenia and southern Moldavia, resulted in the emergence of *Aldeni-Stoicani-Bolgrad* cultural aspect. As a result of these influences, Precucuteni populations began to use elements of painting with white and red, graphite, before and after burning the vessels, moving towards the Cucuteni culture (Ursulescu 1999).

Construction of some homes had been encountered which were partially immersed underground. The material used for the dwellings were wood and clay, which made the structures strong and solid (Marinescu-Bîlcu 1974). Towards the end of the period, besides the homes with earthen floors, the first large houses emerged, with massive clay platforms built on split beams, clay ovens, and appeared the first copper objects (beads, bracelets, pendants). Grinding mills, fairly numerous, were usually oval and very rarely rectangular, large and medium, and the hand mills were made from small tough stones. Flint that was used was mainly black and dark brown in colour (originating from Pruth river area) (Ursulescu 1999).

10.2 Cucuteni Culture

Named after the eponymous settlement of Dealul Cetăţuia, Băiceni village, Cucuteni commune, Iaşi county; it appeared in central and western Moldavia, following phrase III of Precucuteni culture, with the influential impact it had from

Gumelniţa and Petreşti cultures. The discovery of Cucuteni civilisation marked the beginning of archaeological research in eastern Romania, in the first half of the nineteenth century. Undoubtedly, the eponymous settlement from *Cetăţuia* (no. 20) is well known and researched, being located within Valea Oii catchment, this being one of the reasons for which this area was chosen for the study. The folklorist Th. Burda from Iaşi was the first who reported the settlement and made it of high interest to specialists in 1884, followed by systematic research in 1885 (by Butculescu), 1888 (by Beldiceanu), 1909 (by Schmidt).

The Cucuteni culture pottery has several categories: fine, normal with plenty of crushed shells and decorated with comb and lace impressions, belonging to foreign elements, infiltrated from northern and eastern part of Cucuteni-Tripolie cultural complex (Ursulescu 1999). Cucuteni communities pottery is still an emerging field of study, and only a few interdisciplinary studies have revealed the basic pottery techniques (Matau et al. 2012; Buzgar et al. 2013).

The first subphase (A_1) includes settlements with duotone pottery (white and red), being associated with incised ceramics. In subphase A_2, in addition to duotone painting appears the tritone one (white, red, and with a secondary role, black, to delineate the main characteristics outlined with white/red or red/white), and the quantity of incised ceramics decreases. For A_3 subphase, duotone ceramics disappears entirely, being replaced by the tritone one and the incision almost fades. A_4 subphase is making the transition towards A–B phase (Monah 1985).

The painting of **A–B** phase is no longer carried throughout the whole vessel surface. Black colour dominates at the expense of white; the incisions are no longer present on the surface of the vessels. This phase is considered the apex of the Cucuteni civilisation, when they had to adapt fast to the changes of those times (Boghian 2006).

Phase **B**, which is the last one of this culture presents a superior painting; a reddish-yellow background appears on the vessel, with red or black highlights. When some settlements were born, the prehistoric communities had to take into account essential factors: the presence of watercourse in close proximity, favourable land for agriculture, providing defence with minimal effort, and the possibility of exploitation of salt.

There were two different types of settlements: compact (represented by the most of Cucuteni resorts, located on a dominant relief: a promontory, terrace, island, interfluves) and spread out (in open places, without natural or artificial protection). This can also depend on the size of the settlement, the number of dwellings, and how they were arranged: circle (oval, concentric circles), parallel rows or stacked in groups, probably arranged after certain criteria of social and spiritual organisation (Monah 1985; Sorochin 1993; Marinescu-Bîlcu 1997; Boghian 2004).

The Cucuteni settlements were of two types: above ground (the majority) and others partially immersed. They were arranged according to a certain plan and in close proximity to one another. In the centre of the settlement was a dwelling that could represent the place of religious celebrations. It had a split beam, with branches or twigs, then a layer of clay. It is not yet known if these platforms were intentionally burned to be more resistant, or if they were the result of burning the dwellings. The walls were formed from a trellis, stakes and timber, which were covered with clay; and then the roof was covered with reeds. Dwellings above ground had one or more rooms, provided with a clay oven for heating, baking and sometimes objects of worship, in the form of facades and altars, represented by clay zoomorph and anthropomorphic statues, which were thought to be of fertility and fecundity, appearing very expressive and in different poses (Ursulescu 1999).

Within the basin, in some archaeological sites, systematic excavations took place and different analysis from interdisciplinary domains have been made: Cucuteni— *Cetăţuie* (no. 20), Băiceni—*Dâmbul lui Pletosu* (no. 22), Băiceni—*Dâmbu Morii* (no. 17), Băiceni—*Dealul Mănăstirii* (*La Dobrin/Dealul Gosanul*, no. 16), Băiceni —*Hurez* (no. 21), Băiceni—*Silişte* (no. 23), Bălţaţi—*Dealul Mândra* (*La Iaz/Iazul 3*, no. 1), Filiaşi—*Dealul Boghiu* (*Dealul Mare Filiaşi*, no. 5). If at the beginning of the research only 23 sites were known, presently with the help of interdisciplinary research and repeated field trips within the Interdisciplinary Research Platform Arheoinvest, and of a good collaboration between geographers and archaeologists from Romania and France, a number of 26 sites resulted which were integrated into a database with the help of GIS (Asăndulesei 2012; Brigand et al. 2012, 2014). Concerning their periodisation, from the total 26 sites (Table 10.1), a number of 3

Table 10.1 The list of Chalcolithic settlements from the catchment

Site no.	Site name	Location (WGS84 coordinates)		Period/culture	Administrative location (village/commune)
		N	E		
1	Dealul Mândra/la Iaz/Iazul 3	47° 14′ 43″	27° 08′ 7″	Chalcolithic/Cucuteni A$_{3b}$, late Bronze Age/Noua culture	Bălţaţi/Bălţaţi
2	Movila Hârtopeanu	47° 14′ 7″	27° 07′ 27″	Chalcolithic/Cucuteni unknown	Bălţaţi/Bălţaţi
3	Tarlaua Pădurii/Crescătorie 1	47° 15′ 1″	27° 03′ 47″	Chalcolithic/Cucuteni A$_3$	Podişu/Bălţaţi
4	Dealul Oilor/Crescătorie 2	47° 14′ 51″	27° 03′ 35″	Chalcolithic/Cucuteni unknown	Podişu/Bălţaţi
5	Dealul Mare Filiaşi/Dealul Boghiu	47° 15′ 8″	27° 02′ 27″	Chalcolithic/Cucuteni A$_3$	Filiaşi/Bălţaţi
6	SV de Boghiu	47° 14′ 58″	27° 01′ 57″	Chalcolithic/Cucuteni A	Filiaşi/Bălţaţi
7	Dealul Harbuzăriei/V de Dealul Boghiu	47° 15′ 14″	27° 01′ 32″	Chalcolithic/Cucuteni unknown	Filiaşi/Bălţaţi
8	Bejeneasa	47° 16′ 4″	27° 00′ 56″	Chalcolithic/Precucuteni, Cucuteni A, 3–4th centuries	Boureni/Târgu Frumos
9	Dealul Hârtopului	47° 15′ 58″	27° 00′ 5″	Chalcolithic/Cucuteni A–B, B	Boureni/Târgu Frumos
10	Hârtochi/Dealul Hârtop	47° 15′ 44″	26° 59′ 31″	Chalcolithic/Cucuteni unknown	Boureni/Târgu Frumos
11	Bejeneasa I/la Brigadă	47° 17′ 15″	26° 58′ 45″	Chalcolithic/Precucuteni II–III, Starčevo-Cris, Bronze Age/late Hallstatt	Balş/Târgu Frumos
12	Mamelon	47° 17′ 5″	26° 58′ 10″	Chalcolithic/Cucuteni unknown	Balş/Târgu Frumos

(continued)

Table 10.1 (continued)

Site no.	Site name	Location (WGS84 coordinates)		Period/culture	Administrative location (village/commune)
		N	E		
13	Ismiceanu	47° 14′ 7″	27° 03′ 47″	Chalcolithic/Cucuteni unknown	Cucuteni/Cucuteni
14	Târla Luncanului	47° 17′ 15″	26° 56′ 50″	Chalcolithic/Cucuteni A	Cucuteni/Cucuteni
15	Valea Părului III	47° 17′ 52″	26° 57′ 47″	Chalcolithic/Precucuteni II–III	Balş/Târgu Frumos
16	Dealul Mănăstirii/la Dobrin/Dealul Gosanul	47° 14′ 20″	26° 55′ 20″	Chalcolithic/Cucuteni A_3	Băiceni/Cucuteni
17	Dâmbu Morii	47° 14′ 45″	26° 56′ 10″	Chalcolithic/Cucuteni A_2, $A–B_1$, $A–B_2$	Băiceni/Cucuteni
18	La Bazin/fost Gostat	47° 14′ 55″	26° 56′ 0″	Chalcolithic/Cucuteni B, late Bronze Age/Noua I culture, 4th, 10th centuries	Băiceni/Cucuteni
19	Lângă Pod	47° 14′ 7″	27° 03′ 47″	Chalcolithic/Precucuteni, late Bronze Age/Noua I culture	Băiceni/Cucuteni
20	Cetăţuia	47° 17′ 55″	26° 54′ 50″	Chalcolithic/Cucuteni A_2, A_3, $A–B_2$, B_1, B_2	Băiceni/Cucuteni
21	Hurez	47° 18′ 0″	26° 54′ 55″	Chalcolithic/Cucuteni A, 4–3rd century	Băiceni/Cucuteni
22	Dâmbul lui Pletosu	47° 18′ 10″	26° 55′ 20″	Chalcolithic/Cucuteni A, 1–2nd, 4th, 9–10th, 16–17th centuries	Băiceni/Cucuteni
23	Silişte	47° 18′ 10″	26° 55′ 35″	Chalcolithic/Cucuteni unknown, 2–3rd, 8–10th, 15–17th centuries	Băiceni/Cucuteni
24	VSV de vatra satului	47° 18′ 17″	26° 54′ 40″	Chalcolithic/Cucuteni unknown	Băiceni/Todireşti
25	Bârghici	47° 19′ 26″	26° 53′ 15″	Chalcolithic/Cucuteni unknown	Stroeşti/Todireşti
26	Pietrărie	47° 19′ 19″	26° 54′ 13″	Chalcolithic/Cucuteni A, B	Stroeşti/Todireşti

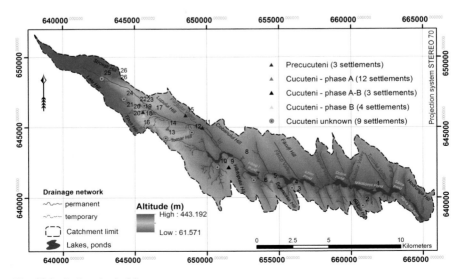

Fig. 10.2 Archaeological inventory

sites belong to the precucuteni period, 13 sites to the Cucuteni A period, 3 sites to the trasition Cucuteni A–B period, 4 sites to the Cucuteni B period and 9 sites have an unspecified period (Fig. 10.2). The ones with unspecified phase were researched only at a surface level being almost impossible to predict a phase in the absence of archaeological excavations.

10.3 The Theory of Island Biogeography

As Candolle (1820) has demonstrated by means of geographical and temporal dynamics in the extent to which biota were influenced by the interaction between the species (development and distribution); achieving a similar analysis for prehistoric populations, based on environmental characteristics is important.

Researcher's attention was particularly directed toward the discovery of the existence of new biota in some remote islands (Darwin, Wallace). Various explanations have reached a sensitive point that the existence of some species in high-mountain areas and in different locations around the globe were unaffected by that of the biblical flood (Willdenow 1972).

Later on Foster, in 1778, following the trip undertaken with the ship H.M.S. Resolution during 1772–1775, was able to observe that the general trend of isolated biota is to be less diverse than those on the mainland; plant diversity increases with the surface of the island, with the number of existing resources, the variety of habitats and energy received from the Sun. The two basic models of the theory are: *the isolation of species* and *species-place relation* (Hutchinson 1959; Hawkins et al. 2003).

These theories regarding the development of biota have in mind the main processes of biogeography: extinction, migration and evolution (Lomolino et al. 2010). Prehistoric populations which, according to regional or local characteristics of the relief, have disappeared, migrated, evolved or either had the ability to adapt. The interaction between species was an important stage in their development process. Island biogeography had its beginnings in the 1950s, when WD Matthew, following several observations of mammals of the northern hemisphere, could see that they were detrimental to those in the southern hemisphere; his conclusion was that a challenging/difficult climate lies behind the strength of a species. Island biogeography is based on two extremely important roles: the cycle of taxons (Wilson 1961) and the balance of species (MacArthur and Connell 1966).

Combining the two concepts, the island biogeography theory resulted. If the catchment is divided into two island areas (upper basin and middle-lower basin), taking into account the dynamics of Cucuteni communities throughout the four stages of evolution (Precucuteni, Cucuteni A, Cucuteni A–B, Cucuteni B), for the determination of the exploited areas, abandonment, stability and development (Asăndulesei 2012; Brigand et al. 2012, 2014), it can be noticed that they abandoned the middle and lower basin (Fig. 10.3). The analysis was made on the basis of the two settlements with a defensive system, *Cetăţuia* (no. 20) and *Dealul*

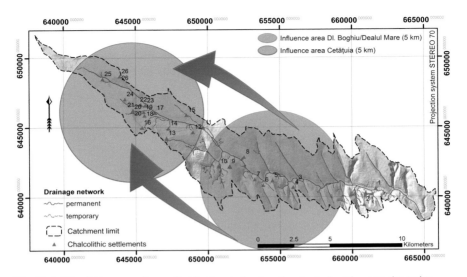

Fig. 10.3 Island biogeography applied in the study of archaeological settlements dynamics

Boghiu/Dealul Mare (no. 5), determined from geophysical prospection, aerial photographs and field investigations.

Instead, in the upper basin, settlements have been positioning themselves around *Cetăţuia* (no. 20), which behaves as a nucleus for the surrounding settlements—one of the fundamentals of insular theory—*the isolation of species*. It is easy to understand why there has been an increase, where natural resources are concentrated: fertile soils suitable for agricultural, high visibility, clay exploitation, forest cover, sulphurous springs, Băiceni and Recea creeks, and the amphitheatre shape-oriented towards south-east, a logical decision in protecting the settlements during the cold season against the Siberian harsh winds (Nicu and Romanescu 2015); an essential strategic position, since it is located at the interfluves between two major river basins, Siret and Pruth—*species-place relation*. The two settlements which do not fall into the two areas of influence, *Dealul Mândra* (no. 1) and *Movila Hârtopeanu* (no. 2), having a peripheral position within the basin, were doomed to fail. The migration process from the upper side of the basin towards the lower side of the basin is analysed and explained by Nicu and Romanescu (2015), with an emphasis to natural risk phenomena, exploitation of new resources, and climatic variations—*species-place relation*.

The observations made are a few hypotheses, the endeavour of realising some (co)relations between the characteristics of the environment and human needs, on the basis of the theory of island biogeography. This can be combated because, in the present case, had not been taken into account the settlements from the neighbouring river basins; it is possible to change the perception and, implicitly, of the result, if a sequential analysis for Bahlui, Jijia or Pruth basins are made. Also, the results are questionable; due to the inadequacy of archaeological information, 9 out of 26 settlements have not yet been determined the phase of evolution.

References

Asăndulesei A (2012) Aplicaţii ale metodelor geografice şi geofizice în cercetarea interdisciplinară a aşezărilor cucuteniene din Moldova. Studii de caz. PhD Dissertation, Univ. "Al. I. Cuza" Iaşi

Bem C (2000) Elemente de cronologie radiocarbon. Ariile culturale Boian-Gumelniţa-Cernavoda I şi Precucuteni-Cucuteni. Cercetări Arheologice XI:337–369, Muzeul Naţional de Istorie, Bucureşti

Boghian D (2004) Comunităţile cucuteniene din bazinul Bahluiului. Editura Universităţii "Ştefan cel Mare" Suceava

Boghian D (2006) Unele consideraţii asupra fazei A-B în contextul civilizaţiei Cucuteniene. Cucuteni 120 – Valori universale: 163–180, Iaşi

Brigand R, Asăndulesei A, Weller O, Cotiugă V (2012) Contribution à l'étude du peuplement Chalcolithique des bassins hydrographiques des Bahluieţ et Valea Oii (dép. Iaşi). Dacia LVI:5–32, Editura Academiei Române

Brigand R, Asăndulesei A, Nicu IC (2014) Autour de la station eponyme de Cucuteni: Paysage et peuplement (Valea Oii, Iasi, Roumanie). Tyragetia SN VIII(XIII)1:89–106

Buzgar N, Apopei AI, Buzatu A (2013) Characterization and source of Cucuteni black pigment (Romania): vibrational spectrometry and XRD study. J Archaeol Sci 40(4):2128–2135. doi:10.1016/j.jas.2012.12.034

Candolle AP (1820) Essai elementaire de Geographie Botanique. Levrault Publishing House

Comşa E (1987) O aşezare Precucuteni din nord-estul Munteniei. Studii şi Cercetări de Istorie Veche şi Arheologie XXXVIII(2):101

Dumitrescu H (1957) Contribuţii la problema originii culturii Precucuteni. Studii şi Cercetări de Istorie Veche şi Arheologie 8(1–4):53–76

Hawkins BA, Field R, Cornell HV, Currie DJ, Guégan J-F, Kaufman DM, Kerr JT, Mittelbach GG, Oberdorff T, O'Brien EM, Porter EE, Turner JRG (2003) Energy, water, and broad-scale geographic patterns of species richness. Ecology 84(12):3105–3117. doi:10.1890/03-8006

Hutchinson GE (1959) Hommage to Santa Rosalia, or why are there so many kinds of animals? American Naturalist 93:145–159

Lomolino MV, Brown JH, Sax DF (2010) Reticulations and reintegration of "A biogeography of the species". In: Losos JB, Ricklefs RE (eds) The theory of island biogeography – revisited, Princeton University Press, Princeton

MacArthur RH, Connell TH (1966) The biology of populations. John Wiley & Sons, New York

Mantu CM (1995) Câteva consideraţii privind cronologia absolută a neoliticului din România, Studii şi cercetări de istorie veche şi arheologie, Institutul de Arheologie "Vasile Pârvan". Bucureşti

Mantu CM (1998) Cultura Cucuteni. Evoluţie, cronologie, legături. Muzeul de Istorie Piatra Neamţ

Matau F, Nica V, Postolache P, Ursachi I, Cotiuga V, Stancu A (2012) Physical study of the Cucuteni pottery technology. J Archaeol Sci 40(2):914–925. doi:10.1016/j.jas.2012.08.021

Marinescu-Bîlcu S (1974) Cultura Precucuteni pe teritoriul României. Bucureşti, pp. 56

Marinescu-Bîlcu S (1997) Consideraţii pe marginea organizării interne a unora dintre aşezările culturilor din Complexul Precucuteni-Cucuteni. Cult Civ Dun Jos XV:165–201

Monah D (1985) Aşezările culturii Cucuteni din România. Iaşi

Nicu IC, Romanescu G (2015) Effect of natural risk factors upon the evolution of Chalcolithic human settlements in Northeastern Romania (Valea Oii watershed). From ancient times dynamics to present day degradation. Z Geomorphol. http://dx.doi.org/10.1127/zfg/2015/0174 (in press)

Petrescu-Dîmboviţa M (1966) Cucuteni. Editura Meridiane. Bucureşti

Petrescu-Dîmboviţa M, Vulpe A (coord.) (2001) Istoria Românilor. Moştenirea timpurilor îndepărtate. Editura Enciclopedică, Bucureşti

Sorochin V (1993) Modalităţi de organizare a aşezărilor complexului cultural Cucuteni-Tripolie. Arheologia Moldovei XVI:69–85, Institutul de Arheologie, Iaşi

Ursulescu N (1999) Începuturile istoriei pe teritoriul României. Casa Editorială Demiurg, Iași

Weide A, Riehl S, Zeidi M, Conrad NJ (2015) Using new morphological criteria to identify domesticated emmer wheat at the aceramic Neolithic site of Chogha Golan (Iran). J Archaeol Sci 57:109–118. doi:10.1016/j.jas.2015.01.013

Willdenow KL (1972) Grundriss de Kräuterkunde zu Vorlesungen (Principles of Botany). Haude und Spender, Berlin

Wilson EO (1961) The nature of the taxon cycle in the Melanesian ant fauna. Am Nat 95 (882):169–193

Chapter 11
Archaeological Sites Affected by Hydrogeomorphological Processes

Abstract The final and the most important chapter of the thesis present three illustrative case studies that are under the direct effect of soil processes: erosion (gully erosion, land sliding) and accumulation (warping). It represents a relatively new field of study, with a few case studies found in the international literature. In order to evaluate de destruction of the three archaeological sites, detailed topographical and geophysical approaches were employed. The monitoring process of the sites is continuing, in order to save important data regarding the endangered cultural heritage of the area.

Keywords Gully erosion · GIS · GPR · Landslide susceptibility model · Rill erosion

The northeast part of the country is well known for its sheepfolds and agricultural use. As such, the area continues to endure intense land degradation by sheep over grazing vegetation and trampling the steep slopes. Erosion processes are accelerating under the conditions of a clay substrate, torrential rains during the warm season and steep slopes. The erosion processes occurring in the basin have been previously mentioned in several limited works, historical documents (Corsi et al. 2009). "*Această coastă este mult fragmentată, datorită eroziunii apelor, care au ferestruit până la adâncime podişul prin valea Băicenilor, valea Cârjoaiei...*" (Bucur and Barbu 1954)—"*This cuesta is very fragmented because of water erosion, which deeply cut the plateau across the Băiceni valley, Cârjoaia valley...*". "*Cele mai frecvente forme de eroziune torenţială, în diferite stadii de evoluţie, se întâlnesc către periferia Câmpiei Moldovei, în zona localităţilor Tomeşti, Voineşti, ... şi pe versanţii cu caracter de coastă ai Jijiei, Bahluiului, Bahluieţului, Miletinului, Sitnei, Başăului, Văii Oilor etc.*" (Băcăuanu 1967)—"*The most frequent forms of torrential erosion, in various states of evolution, can be found towards the periphery of the Moldavian Plain, around the settlements of Tomeşti, Voineşti, ... and on the versants front of cuesta character of the Jijia, Bahluieţ, Miletin, Sitna, Başău, Oii valley, etc.*". "*Un tip aparte de eroziune liniară se întâlneşte pe versanţii care au deschise la zi orizonturi de marne sarmaţiene cu un grad mai ridicat de salinizare. La Dumeşti-Iaşi..., pe versantul sudic al Văii*

© The Author(s) 2016
I.C. Nicu, *Hydrogeomorphic Risk Analysis Affecting Chalcolithic Archaeological Sites from Valea Oii (Bahlui) Watershed, Northeastern Romania*,
SpringerBriefs in Earth System Sciences, DOI 10.1007/978-3-319-25709-9_11

Oilor..." (Băcăuanu 1967)—"*A particular type of linear erosion can be found on the slopes that have opened horizons of Sarmatian marls with high degrees of salinisation. At Dumeşti-Iaşi,..., on the southern part of the Oii valley*...".

The basin is not characterised by a large development of erosion processes. After analysing the affected surfaces from erosion extracted from the pedological maps at 1:10,000 scale, it resulted in more than half of the basins surface is not eroded (40 %), poorly eroded (31 %) or moderately eroded (21 %); these areas are mainly developing on the right side of the basin and the structural plateaus (in the upper basin, in the plateau area). The processes are especially concentrated in the upper basin, at the contact between the plain and plateau. The area of Moldavian Plateau is known to be highly susceptible to landslides (Surdeanu 1998).

At an international level, the research regarding the degradation of archaeological sites by landslides are numerous and spread throughout different parts of the world, because considerable efforts are being made for saving and conserving the international cultural heritage (Canuti et al. 2000; Christaras et al. 2002; Grossi et al. 2007; Sdao and Simeone 2007; Alexakis and Sarris 2010; Eeckhaut et al. 2010; Nikolova et al. 2012; Tarragüel 2012; Gaynullin et al. 2014). The Romanian literature is lacking in studies of this kind (Mara and Vlad 2008; Micle 2014; Nicu and Romanescu 2015).

From the geomorphological processes, the most frequent are the landslides, the study area having high potential, high probability and medium susceptibility to landslides (Bălteanu et al. 2010). These occupy approximately 110 ha (9.8 % from the total surface of the basin; the landslides were extracted out of the orthophotoplans at a 1:5000 scale—2005 edition, combined with direct field observations and surveys with a total station and a GPS), spread mainly on the right side of the basin.

For producing Landslide Susceptibility Model (LSM), the *Raster Calculator* function of the ArcGIS software suite was used, for the following thematic layers: lithology, slope, parent material, soil classes, precipitations and land use. The vectors of each thematic layer were subsequently transformed to Raster (*Feature to Raster*), each class having been assigned a susceptibility class (low—1, moderate—2, high—3, very high—4). For each vector layer, a value has been given in the final equation of susceptibility index (geology—15 %, parent material—10 %, land use —15 %, slope—35 %, soil classes—10 %, precipitations—15 %), according to the available spatial data and to values encountered in the international literature on calculating the susceptibility to landslide using GIS (Wachal and Hudak 2000; Thiery et al. 2007; Conoscenti et al. 2008; Klose et al. 2014). From the LSM map (Fig. 11.1), it can be observed that the class with low susceptibility occupies 23.4 %, with moderate susceptibility 57.8 %, with high susceptibility 10.5 %, and very high susceptibility 8.3 %. Accordingly, the basin falls within the limits set by Bălteanu et al. (2010), after producing the model of landslide susceptibility for the entire country. For validating the empirical model, the vectorial layer was overlapped by the mapped landslides, from which it can be observed that the latter occur in areas with high and very high susceptibility, which indicates that the model is viable; similarly, the areas with very high and high susceptibility correspond to the landslides' depletion areas, to narrow valleys with steep slopes on the left side of

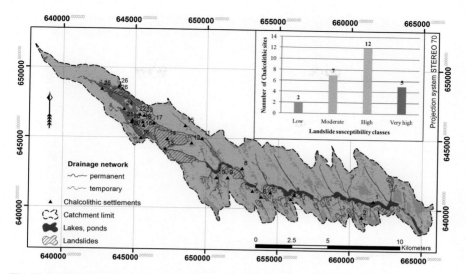

Fig. 11.1 The computed LSM (landslide susceptibility model) of the catchment

the basin and affected by processes of soil salinisation. The most susceptible area is found in the upper basin, at the contact between the plain and the plateau. From the total of 26 Chalcolithic sites, 2 are located in areas with low susceptibility, 7 in areas with moderate susceptibility, 12 in areas with high susceptibility, and 5 in areas with very high susceptibility. As it can be observed, more than 65 % of the sites are located in areas with high and very high susceptibility, raising the chances of degradation in the future.

The landslides with the largest surface (160 ha) are found in the upper basin, in the area of Cucuteni, Băiceni and Stroieşti villages, which are very old and stabilised, as a result from past deforestation. For example, the stabilised landslide is affecting *Cucuteni-Cetăţuie* (no. 20) settlement and a large area of Băiceni village, which is located on the huge deluvium, and is at least Pleistocene age.

Landslides with smaller surfaces, that are recent and generally stabilised, were formed in the middle and lower basin, conditioned by the steep slope (5°–7°), lack of vegetation (sheep farming being one of the basic occupation of people from the catchment), the landuse (grazing), and by building ponds on the main course. These, by their presence contribute to the undermining, instability of the slopes and triggering landslides. Measures to stop the process are found in the northwestern part of Oilor Hill, near Podişu pond, where gabions of stones were placed to stop the advance of deluvium into the road between Podişu village and Târgu Frumos town (Fig. 11.2a). In total, a number of 12 archaeological sites (Nicu and Romanescu 2015) are affected by this process (Table 11.1). In addition to damage and destruction, evidence that will never be recovered; in most cases, along with deluvium archaeological remains are carried (ceramic pieces, flints) down the slopes (such as landslide that affects *Dealul Boghiu*, which is a typical case study, Fig. 11.2b).

Fig. 11.2 a The gabions of stones placement at the bottom of the landslide; **b** pottery pieces carried out down the slopes of the landslide

Gully erosion occupies an area of approximately 20 ha (0.2 % of the total catchment surface); the total number of gullies within the catchment is 25. Gullies are found exclusively on the right side of the basin, with steep slopes and deluvial coasts.

Table 11.1   The outlook of the Chalcolithic sites affected by hydrogeomorphic processes

Hydrogeomorphological processes		Name of affected site
Gully erosion		*Cetăţuia, Dealul Mănăstirii (la Dobrin/Dealul Gosanul), Dealul Boghiu (Dealul Mare/Filiaşi)*
Landslides	Active	*Dealul Boghiu (Dealul Mare/Filiaşi), Movila Hârtopeanu*
	Stabilised	*Tarlaua Pădurii, Dealul Oilor, SV de Boghiu, Bejeneasa, Ismiceanu, Târla Luncanului, Dâmbu Morii, La Bazin, Lângă Pod*
Alluvial/warping and back water process		*Dealul Mândra (la Iaz/Iazul 3)*
Complex (landslides and gully erosion)		*Cetăţuia, Dealul Boghiu (Dealul Mare/Filiaşi)*

Affected by this process are the following settlements: *Cetăţuia* (no. 20), *Dealul Mănăstirii (la Dobrin/Dealul Gosanul)* (no. 16), *Dealul Boghiu (Dealul Mare/Filiaşi)* (no. 5), which are one of the most important settlements in the basin. This is the reason why the gully that is affecting the site *Dealul Mănăstirii (la Dobrin/Dealul Gosanul)* (no. 16) was chosen as a case study within this chapter. Worldwide, there are only a few studies about gully erosion affecting archaeological sites (MacDonald 1990; Pederson et al. 2006; Romanescu et al. 2012; Romanescu and Nicu 2014; Nicu and Romanescu 2015).

Sedimentation and erosion represent active processes which, in time, have contributed to the modelling of the landscape. From a geographical point of view, they provoke imbalances in basin systems. Human activities accelerate the sedimentation and erosion processes. The thickness of the sediments is proportional with the mechanic characteristics of the soil, properties of the geological substrate, capacity of water transportation, morphometric characteristics of the basin (slope, altitude), the degree of vegetation coverage, land use, and the intensity of human activities within the basin. In a general way, besides the in situ negative effects of translocation of soil particles, erosion damages the agricultural lands by reducing soil fertility and transporting sediments, affecting water quality. Concerning the present case study, the sedimentation was present before the building of the dam and after filling it with water. Sedimentation has also contributed in covering potential archaeological traces. Sedimentation processes can have a positive effect by conserving archaeological settlements (if they existed).

The building of dams affect the natural evolution of the valley in two ways: stopping natural fluxes of sediments and sediment load upstream. Another result from the building of dams is the water overload that can cause floods. Sometimes dams are built with the aim of stopping negative effects associated with maximum flow; they can also be a triggering factor in producing of such phenomena.

Streams have an action of erosion, transport and deposition of sediments. Erosion produced by water streams has the natural tendency of reducing the angle of inclination and achieving equilibrium profile. The action is influenced by the transport capacity and is proportional with the stream velocity—erosion and transport on the upper course, transport and deposition on the middle course and deposition on the lower course. Earthworks on the water streams have a local influence on water flow channel; earthwork on main course leads to steep slope and also the increase of water flow (Julien 2010).

The alluvial/warping/fluvial erosion affects the settlement *La Iaz (Iazul 3/Dealul Mândra)* (no. 1), located on the left bank of the valley, close to the main course. Prior to the dams construction, the site was affected by alluvial processes, which ceased after the dam was built. Following the construction of Sârca dam, a big part of the site is underwater, being under the continuous level oscillations.

11.1   Archaeological Sites Affected by Gully Erosion (Case Study)

Dealul Mănăstirii/la Dobrin/Dealul Gosanul, site located in the upper part of the basin (geographical coordinates WGS 84: 47° 17′ 20″ lat. N, 26° 55′ 20″ long. E; STEREO 70: $X = 645383.084$, $Y = 645024.074$) in the area of transition between the Moldavian Plain and the Suceava Plateau, on the Cucuteni commune territory, area known in the geomorphological literature as Coasta Dealul Mare Hârlău, on the south-east border of the Laiu plateau. The geomorphological processes have a pronounced character *"this coast is highly fragmented due to water erosion, which dug deep in the Baicenilor valley plateau…"*. The gully erosion process has been recorded in literature *"torrential gullying is characteristic on Dealul lui Voda and is frequent enough in Deleni-Hârlau region, Cucuteni-Baiceni region, and due to the regressive retreat of the torrents in the contact area towards the lowlands…"* (Bucur and Barbu 1954).

Within the catchment there are a number of 25 gullies, which occupy 1 % of the basin area. They are found exclusively on the right side of the catchment. The majority of gullies do not affect the economic activities, being located on pastures, and only nine affect archaeological sites. This gully was chosen for its evolution in affecting archaeological sites of Neolithic age *Dealul Mănăstirii/la Dobrin/Dealul Gosanul* (no. 16), and the Geto-Dacian settlement *Mlada*. The Neolithic settlement belonging to the Cucuteni A$_3$ period is in a fast process of degradation; after the archaeological research undergone by Chirica V., Popovici R., Iconomu C. in 1979 and by Petrescu-Dâmbovița M. in 1981, settlement remains, ceramics, a fragment of an anthropomorphic idol, and skeletal remains were discovered (Chirica and Tanasachi 1984). Regarding the Geto-Dacian site Mlada, this was researched by Laszlo A., excavations being made between 1964 and 1966, five archaeological sections were dug, revealing two levels of habitation, holes, and many archaeological pottery remains.

Interdisciplinary research was set in the fall of 2008 and underwent with the state of the art equipment ever since (Mihu-Pintilie and Romanescu 2011; Nicu and Romanescu 2011; Romanescu et al. 2012; Nicu and Mihu-Pintilie 2012; Romanescu and Nicu 2014; Nicu and Romanescu 2015), being researched under

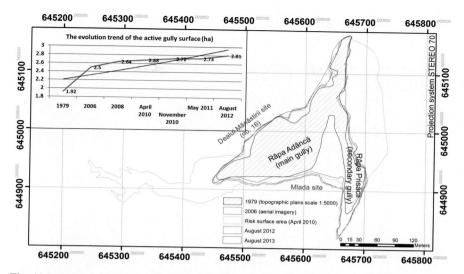

Fig. 11.3 Băiceni-Cucuteni gully dynamics highlighted by systematic topographic surveys

geomorphologic aspect in 2010 (Chiriac 2010). The gully has a high level of importance as it was set through the interdisciplinary research made, continuing to be under our close observation, being a typical case study in the field of cultural heritage protection and land degradation with effects on archaeological sites. During the monitoring process of the gully since 2008, in addition to surveys made with total station Leica TCR1201, GPS System 1200 and producing maps with gully dynamics (Fig. 11.3), a series of photos were made, which highlights the dynamics of the upper part located between Râpa Adâncă (the main gully) and Râpa Prisăcii (the secondary gully) (Fig. 11.4).

From the geomorphological point of view, the gully has the following typology: after location in the basin: the slope; by cross-section shape: "V" shape; after the development cycle: perennial; longitudinal profile form: continuous; as the plan configuration: dendritic (with two points of origin) (Romanescu et al. 2012); size: very big (with depths of 20–25 m); after catchment size: small; after surface: big; after gully process intensity: middle.

The mean gully annual advance for Moldavian Plateau is 1.5 m/year, calculated by empirical models (Radoane et al. 1995), this gully has a faster evolution; after consulting and analysing the cartographic background of the gully dynamics from 1979 until present, the most active part of the gully had an average annual regression of 3.5 m/year, reported to a period of 34 years, respectively, 2.8 m/year, reported to a period of 7 years (since the monitoring process of the gully started).

The regression rate of the gully can vary from year to year, according to the meteorological conditions and the anthropical interventions; inside it many other processes are being developed such as geomorphological processes (landslide) noted fact by the archaeologists in literature "*the ESE part of the resort is*

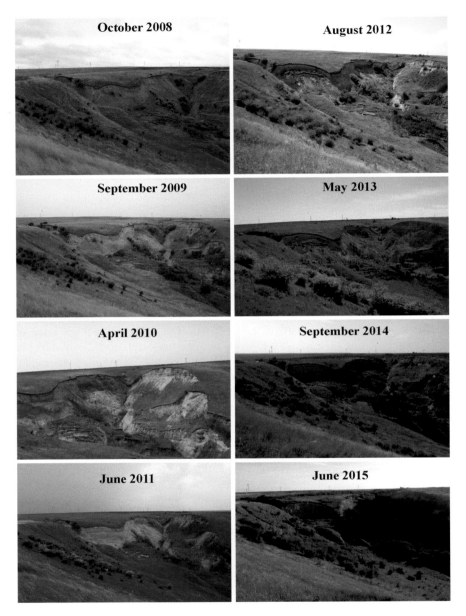

Fig. 11.4 Băiceni-Cucuteni gully dynamics highlighted by successive photographs

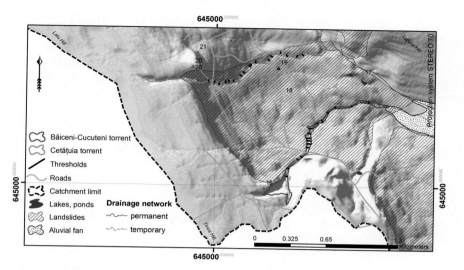

Fig. 11.5 The location of the thresholds inside the two torrents from the upper catchment

destroying due to landslides" (Chirica and Tanasachi 1984). The Baiceni-Muzeu Cucuteni gully is characterised through a low hydrological activity almost all year; even if it has a small drainage area, when torrential rainfall occurs, it presents a potential hydrological risk of maximum flow (Nicu and Mihu-Pintilie 2012). Other important factors which enhance the erosion process are caused by the variation of the hydrostatic level which ultimately leads to collapse, trenches from World War II that facilitate the infiltration of water in the soil. Considering the seismic activity intensification in Romania has been since 2013, according to the provisions of the P100-92 Normative for anti-seismic design of building and other constructions, the area falls into the "C" seismic category, characterised by:

Ks   seismicity coefficient = 0.20;
Tc   corner period = 0.7 s;
Msk   macro-seismic intensity = VII

The calculus insurance. According to STAS 4273–83, with the changes brought through M–SR 6/83 and 2/87, concerning the assignment into classes of importance of hydrotechnical works, respectively, STAS 5576/88 for assigning into classes of importance the works for controlling torrents, the designed works in terms of functional significance are principal, and in terms of uselife they are permanent (definitive). From the point of view of economic importance, according to the sought objectives, the works fall into category IV of importance. The category of importance of the construction determined according to the "Methodology for establishing the category of importance of the constructions", sanctioned by MLPAT Order no. 31/N/1995, is C—normal (SC Wareso Prod SRL 2011).

Anti-erosion measures consisted in planting, in the spring of 2011, of Black Locust (*Robinia pseudoacacia*) saplings, fences of fascine were placed on the right side of the secondary gully, and finally, during the summer of 2014, the construction of concrete thresholds along the main thalweg were started; the management plan indicated a construction total of 16 thresholds for this gully. For the torrent Cetățuia, placed 1 km up to the north, a number of 17 thresholds (Fig. 11.5).

11.2   Archaeological Sites Affected by Landslides (Case Study)

Dealul Boghiu/Dealul Mare approximately located in the middle part of the basin (geographical coordinates WGS 84: 47° 15′ 7.2″ lat. N, 27° 02′ 27″ long. E; STEREO 70: X = 654471.048, Y = 641163.201); the site of Cucuteni A_3 period is situated roughly 600 m South of the Filiași village and approximately 2 km West of the Podișu village (both belonging to the Bălțați commune, Iași county), on a front of cuesta with an altitude of 185 m orientation north. The settlement with a hectare of 1.5 was discovered and researched for the first time by Orest Tafrali in the fall of 1935, which, along with Emil Condurachi and Victor Manoliu have undergone a series of archaeological surveys with unexpected results. Other research followed by Berlescu N. in 1955, Boghian D. and Mihai C. between 1984 and 1986.

Tafrali locates the settlement "*la vreo opt kilometri la Nord de Târgu Frumos, între acest oraș și satul Cotnari, se află un deal, care domină câmpia înconjurătoare. Este punctual cel mai ridicat din această regiune. El face față înălțimilor de la Cotnari depărtate cu circa zece kilometri. Locuitorii numesc acest deal Boghiu*" (Tafrali 1936)—"*about eight kilometres north from Târgu Frumos city; between this city and Cotnari village, there is a hill that dominates the surrounding plain. It is the highest point in the region. There is direct visibility between this hill and Cotnari Hill, located at about 10 km north. The inhabitants named this hill Boghiu Hill*". This feature of the landscape was favoured, by the location of the settlement, a place with good visibility with the site from *Dealul Cetățuia* (no. 20) (defensive purpose); other advantages are the proximity towards the main river, the existence of two springs near the coast, a vast plateau towards the south with fertile soils where they could practice agriculture, one of their main activities.

Among the most significant archaeological finds were a large number of houses, polished stone axes, three-colour painted ceramics, an *en violon* idol, soldering platform, the most interesting find being the miniature clay chair (Boghian 2004; Văleanu 2003). Latest research (archaeological topography, detailed surveying, non-invasive investigations, aerial photos) was obtained by the staff of the Interdisciplinary Arheoinvest Platform from Alexandru Ioan Cuza University of Iasi in 2010, 2011, 2012, 2013 and 2014.

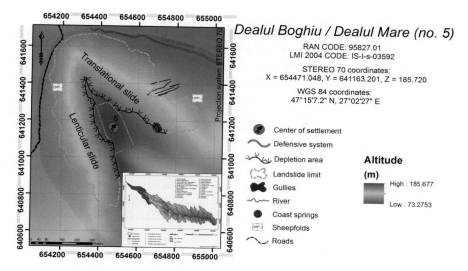

Fig. 11.6 Detailed topographic survey of *Dealul Boghiu/Dealul Mare*

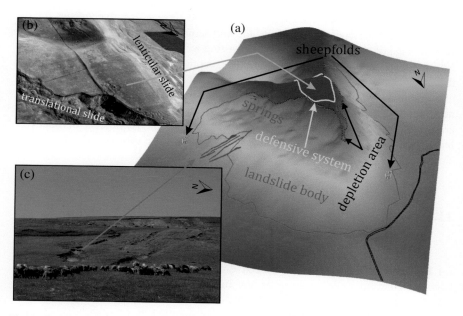

Fig. 11.7 a 3D view of *Dealul Boghiu/Dealul Mare*; **b** Aerial photo (by A. Asăndulesei); **c** Gully erosion on the eastern slopes

The topographic survey was done over two days in June 2011; approximately 2000 points were measured with the help of the total station, in STEREO 70 coordinate system, completed with roughly another 1000 points measured with the GPS in May 2013 (Fig. 11.6). By using the GIS extension (ArcScene), a 3D model of the settlement was possible with all the cartographic elements included (Fig. 11.7a).

This site is intensively being affected (including the anthropogenic fortification system) by an active landslide, with a surface of 32 ha (Fig. 11.7b), by gullying processes in the east side of the slope (Fig. 11.7c), also by anthropic activities (overgrazing, diggings undergone by local authorities without the archaeologists agreement, a big hole for clay exploitation, with a surface of 800 m² and depth of approximately 8 and 10 m in diameter, with a disregard in violation of Law no. 43/30 January 2000 regarding the protection of archaeological cultural heritage). The 32 ha landslide falls in the category of a very large landslide (Cornforth 2005). Analysed from a typological point of view, the left side of the landslide is of lenticular type, and the right side is of translational type.

Unfortunately for this site, no anti-erosion control measures were ever made in order to ameliorate or stop the geomorphological processes; these, along with anthropic ones (overgrazing) are factors that lead to the acceleration of soil degradation. Erosion control measures must be implemented as soon as possible to stop or at least to mitigate these processes which have a negative impact for the archaeological heritage. In Romania, the archaeological heritage is defined as the "assemblage of archaeological goods consisting in: archaeological sites listed in the National Archaeological Registry, with the exception of those destroyed or lost, and the sites classed into the List of Historical Monuments, found above ground, underground or underwater, comprising archaeological vestiges: settlements, necropolis, structures, buildings, groups of buildings, as well as lands with repeated archaeological potential, as defined by law; mobile goods, objects or traces of human activities, alongside the land on which they were discovered" (Lazăr and Condruz 2007).

11.3   Archaeological Sites Affected by Alluvial/Warping and Back Water Processes (Case Study)

La Iaz/Iazul 3/Dealul Mândra placed on the interfluve area with gentle slopes between Turcului Valley to the East and Babelor Valley to the West (geographical coordinates WGS 84: 47° 14′ 43″ lat. N, 27° 08′ 80″ long. E; STEREO 70: $X = 661630.029$, $Y = 640579.428$), in the point called "La Gorciu", at approximately 2.5 km NNE from the Bălțați village, on the left side of the valley.

A detailed topographic survey was undertaken in September 2012, highlighting the settlement topography, a cartographic background for the future non-invasive prospection's of the submerged settlement and to estimate the area. Archaeological field research was made by the staff of Interdisciplinary Research Department Arheoinvest Platform (Nicu Ionuț Cristi, Asăndulesei Andrei, Brigand Robin, Cotiugă Vasile), finding a diverse range of archaeological artifacts (axes, phalusses); on the entire surface of the site an abundant of archaeological remains still exist, especially on the Southern part where the settlement continues underwater.

In the case of this settlement, although it is contemporaneous with those from *Dealul Mănăstirii* (no. 16) and *Dealul Boghiu/Dealul Mare* (no. 5) placed on a high relief, is located at an altitude of 73 m; this factor was not a crucial one in the placement of the site, although having direct visibility with the settlement *Dealul Boghiu/Dealul Mare* (no. 5). It can be stated that the settlement's close proximity towards the main river course and the existence of three springs (one is presently intercepted) (Nicu et al. 2012), visible only when the water level from Sârca pond is low, were the main reasons for this dwelling's placement. Geomorphologically speaking, the site is located on the cuesta reverse from the left side of the catchment with gentle slopes of 3°–5°, with fertile soils, for agricultural purposes.

Before building of the dams along the basin between 1961 and 1962, the settlement was affected by sedimentation processes; with the building of dams, the process was stopped. The anthropic activities which affect the settlement are present within the entire area, agriculture having the highest impact. A significant part of the site was destroyed between 1961 and 1962 when the dams were built.

A number of four dwellings are visible on the south-eastern side of the site, which are in a continuous degradation process because of water level oscillation of Sârca pond. That is why GPR surveys were undertaken in January 2013 on the frozen Sârca pond surface and in close proximity to the site. The purpose for this was to find other possible dwellings that could potentially be under the water and also to identify the sediment thickness from the pond. GPR (Malå X3 M) with a shielded antenna of 250 MHz was used, being the most appropriate and practical one for medium and shallow depths (Fig. 11.8a). As for a precise location of the GPR profiles, GPS Leica 1200 was used (Fig. 11.8b). In total, a number of 12 profiles were undertaken.

Once the GPR measurements were done, the depth of the pond was determined and identified possible archaeological anomalies. Profiles had shown the thickness of sedimentation at the bottom of the pond; after the data processing was achieved with the help of RadExplorer 1.41 (Malå Geoscience, Sweden), profile no. 876 and no. 878 were the ones which caught our attention and interest.

For depth calibration during the data processing, the standard value for relative dielectric permittivity (Conyers 2004) for water 0.04 m/ns was used (the values

Fig. 11.8 a GPR surveys with 250 MHz shielded antenna; **b** GPS used to georeference GPR profiles

coincide for water in liquid form but also in solid-ice form, like in the present case). Therefore, with the help of 250 MHz antenna, after data calibration, a maximum penetration depth of 3.8 m was reached. The 876 profile has a length of 80 m, on the SW-NE direction, parallel with the bank of the Sârca pond. Covering the entire

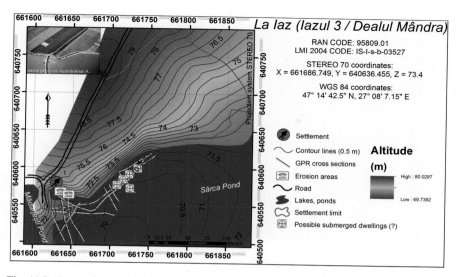

Fig. 11.9 General map of the settlement showing the detailed topography, GPR profiles location and the assumptive dwellings

length of the profile there were a number of four anomalies identified, starting from 12 m until 60 m, which can be associated with possible traces of archaeological dwellings.

After the correction of the altitudinal points marked with the GPS resulted a flat surface of about 0.13 ha (1325.12 m²), favouring the construction of 4–5 houses; we state this fact due to the 5 possible submersed settlements found in continuation of the other 4 settlements visible on the shore (Fig. 11.9). Regarding the sediment thickness, it starts from 0.5 m close to the shore, until 3 m at a distance of about 30 m from shore. Limitations of the 250 MHz antenna are obvious, making it unsuitable for reaching deep depths; in this case the usage of the 100 MHz is required (this will be accomplished in the future for widening the research related with the warping of all the ponds within the basin) which, in ideal conditions can reach depths of 20–25 m. What is found under the sediments is, most likely the valley conformation before warping. In this case, the GPR represents the best non-destructive method for determining the sediment thickness, but also high resolution images (Bristow 2003).

To make the switch to the next observations related to the following profile— 878 (Fig. 11.10) and 876 (Fig. 11.11) they intersect each other at 28 m. Profile 878 has a 30 m length and was configured on the NW-SE direction, approximately perpendicular with the ponds shore, and also on profile 876. Some anomalies can be

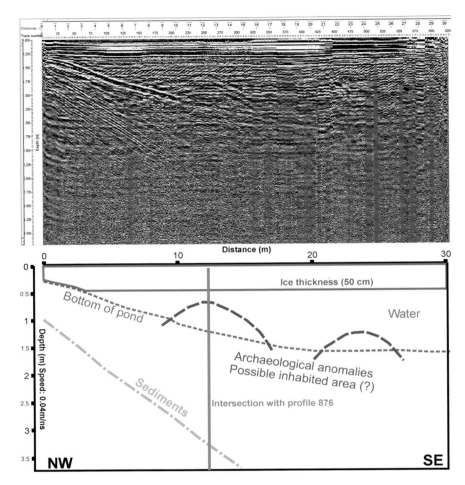

Fig. 11.10 Detail with GPR profile no. 878 and the interpretation of the data

observed which can be linked with traces of settlements. The scans were done in the lower part of the valley, where the sediments are deposited, at least at a theoretical level. Ideally these scans should be accompanied by a dating method, in order to have a better overview of the valley evolution. Being a basin in the plains area, the water transport capacity is not as high as it would be in the case of a basin in a piedmont or mountain area, the fact being it should not be neglected.

Besides these natural processes, the NNE part of the settlement is also affected by agricultural practices in an intensive manner, where the Chernisols are present

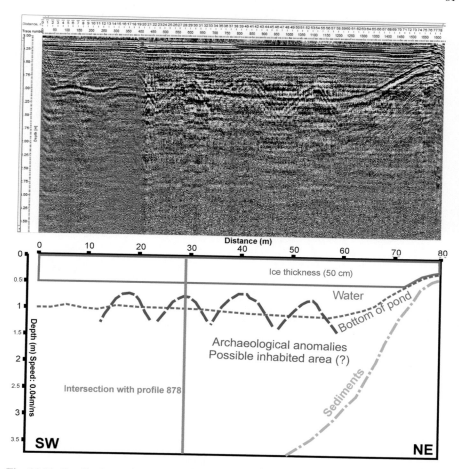

Fig. 11.11 Detail with GPR profile no. 876 and the interpretation of the data

with a high fertility and productivity. In the case of both profiles, the results are not absolute, they can differ with the possibility of a significant decrease in the water level of Sârca pond; this can make it more accessible for locating the underwater dwellings and eventually archaeological excavations can be undertaken. Another method which could confirm the findings is coring the submerged area, where stratification can be observed in detail. Finally, being able to overlap the GPR profiles (Fig. 11.12) offering a clearer 3D perspective for what is taking place under the Sârca pond water level.

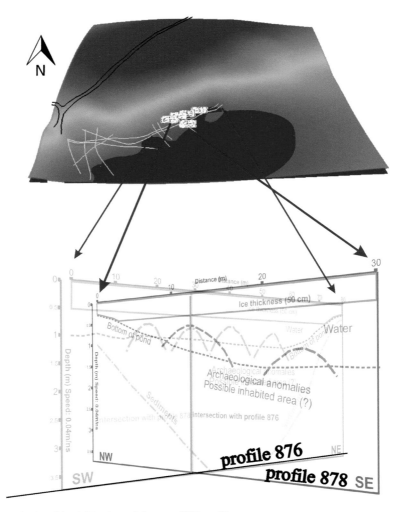

Fig. 11.12 Combined 3D view of the two GPR profiles

References

Alexakis D, Sarris A (2010) environmental and human risk assessment of the prehistoric and historic archaeological sites of Western Crete (Greece) with the use of GIS, remote sensing, fuzzy logic and neural networks. In: Ioannides M (ed) EuroMed. LNCS, vol 6436, pp 332–342

Băcăuanu V (1967) Cercetări geomorfologice asupra Câmpiei Moldovei (extras). An. şt. ale Univ. "Al. I. Cuza" (serie nouă), sect. II, tom XIII, Iaşi

Bălteanu D, Chendeş V, Sima M, Enciu P (2010) A country-wide spatial assessemt of landslide susceptibility in Romania. Geomorphology 124:102–112. doi:10.1016/j.geomorph.2010.03.005

Boghian D (2004) Comunitățile cucuteniene din bazinul Bahluiului. Editura Universității "Ștefan cel Mare" Suceava

Bristow CS, Jol HM (eds) (2003) Ground penetrating radar in sediments. Geological Society, London (Special Publications)

Bucur N, Barbu N (1954) Complexul de condiții fizico-geografice din ≪Coasta Dealul Mare – Hîrlău≫. Probleme de Geografie, vol. I, Editura Academiei RPR

Canuti P, Casagli N, Catani F, Fanti R (2000) Hydrogeological hazard and risk in archaeological sites: some case studies in Italy. J Cult Herit 1(2):117–125. doi:10.1016/S1296-2074(00)00158-8

Chiriac C (2010) Geo-morphological aspects from Cotnari Vineyard area. Lucr. Sem. Geogr. "D. Cantemir" 30:63–71

Chirica V, Tanasachi M (1984) Repertoriul arheologic al județului Iași. vol. I, Iași

Christaras B, Mariolakos I, Dimitriou A, Moraiti E, Mariolakos D (2002) Slope instability at Olympia archaeological site, in S. Greece. In: International symposium of UNESCO "landslides risk mitigation and protection of cultural and natural heritage", Kyoto, pp 339–342

Conoscenti C, Di Maggio C, Rotigliano E (2008) GIS analysis to asses landslide susceptibility in a fluvial basin of NW Sicily (Italy). Geomorphology 94(3–4):325–339. doi:10.1016/j.geomorph.2006.10.039

Conyers LB (2004) Ground-penetrating radar for archaeology. Alta Mira Press, Walnut Creek

Cornforth DH (2005) Landslides in practice—investigations, analysis, and remedial/preventative options in soils. Wiley, New York

Corsi C, De Dapper M, Vermeulen F (2009) River bed changing in the lower Potenza Valley (mid-Adriatic Italy). A Geo-archaeological approach to historical documents. Z Geomorphol N.F 53(1):83–98. doi:10.1127/0372-8854/2009/0053s1-0083

Eeckhaut MVD, Poesen J, Vandekerckhove L, Gils MV, Rompaey AV (2010) Human-environment interactions in residential areas susceptible to landsliding: the Flemish Ardennes case study. Royal Geogr Soc 42(3):339–358

Gaynullin II, Sitdikov AG, Usmanov BM (2014) Abrasion processes of Kuibyshev reservoir as a factor of destruction of archaeological site Ostolopovo (Tatarstan, Russia). Adv Environ Biol 8 (4):1027–1030

Grossi CM, Brimblecombe P, Harris I (2007) Predicting long term freeze-thaw risks on Europe built heritage and archaeological sites in a changing climate. Sci Total Environ 377:273–281. doi:10.1016/j.scitotenv.2007.02.014

Julien PY (2010) Erosion and sedimentation, 2nd edn. Cambridge University Press, Cambridge

Klose M, Gruber D, Bodo D, Gerhard G (2014) Spatial databases and GIS as tools for regional landslide susceptibility modeling. Z Geomorphol 58(1):1–36

Lazăr A, Condruz A (2007) Corpus Juris Patrimonii. Patrimoniul Cultural Național. Lumina Lex, București

Mara S, Vlad SN (2008) Positive effects of natural hazards on cultural heritage in Romania. Geogr Fis Dinam Quat 31:181–186

MacDonald A (1990) Surface erosion and disturbance at archaeological sites: implications for site preservation. Miscellaneous Paper EL-90-6, US Army Engineer Waterways Experiment Station, Vicksburg, MS

Micle D (2014) Archaeological heritage between natural hazard and anthropic destruction: the negative impact of social non-involvement in the protection of archaeological sites. Procedia Soc Behav Sci 163:269–278

Mihu-Pintilie A, Romanescu G (2011) Determining the potential hydrological risk associated to maximum flow in small hydrological sub-basin with torrential character of the river Bahlui. PESD 5(2):255–266

Nicu IC, Romanescu G (2011) Determination of ground-water level by using modern non-distructive methods (GPR technology). Air and water components of the environment:441–448, Cluj-Napoca

Nicu IC, Mihu-Pintilie A (2012) Hydrogeomorphological risk analysis models in experimental river basins. Case study: Băiceni-Cucuteni Museum gully (Valea Oii watershed). Lucr. Sem. Geogr. "D. Cantemir" 34:15–22

Nicu IC, Asăndulesei A, Brigand R, Cotiugă V, Romanescu G, Boghian D (2012) Integrating geographical and archaeological data in the Romanian Chalcolithic. Case study: Cucuteni settlements from Valea Oii (Sheep Valley—Bahlui) watershed. Geomorphic processes and geoarchaeology. From landscape archaeology to archaeotourism, Moscova—Smolensk, ≪Universum≫:204–207

Nicu IC, Romanescu G (2015) Effect of natural risk factors upon the evolution of Chalcolithic human settlements in Northeastern Romania (Valea Oii watershed). From ancient times dynamics to present day degradation. Z Geomorphol. doi:http://dx.doi.org/10.1127/zfg/2015/0174 (in press)

Nikolova M, Nedkov S, Nikolov V (2012) Risk from natural hazards for the archaeological sites along bulgarian Danube bank. In: European SCGIS conference, pp 90–96

Pederson JL, Petersen PA, Dierker JL (2006) Gullying and erosion control at archaeological sites in Grand Canyon, Arizona. Earth Surf Proc Land 31:507–525. doi:10.1002/esp.1286

Radoane M, Ichim I, Radoane N (1995) Gully distribution and development in Moldavia, Romania. Catena 24(2):127–146. doi:10.1016/0341-8162(95)00023-L

Romanescu G, Cotiugă V, Asăndulesei A, Stoleriu CC (2012) Use of the 3-D scanner in mapping and monitoring the dynamic degradation of soils: case study of the Cucuteni-Baiceni Gully on the Moldavian Plateau (Romania). Hydrol Earth Syst Sci 16:953–966. doi:10.5194/hess-16-953-2012

Romanescu G, Nicu IC (2014) Risk maps for gully erosion affecting archaeological sites in Moldavia, Romania. Z Geomorphol NF 58(4):509–523. doi:10.1127/0372-8854/2014/0133

Sdao F, Simeone V (2007) Mass movements affecting Goddess Mefitis sanctuary in Rossano di Vaglio (Basilicata, southern Italy). J Cult Herit 8(1):77–80. doi:10.1016/j.culher.2006.10.004

Surdeanu V (1998) Geografia terenurilor degradate I. Alunecări de teren. Presa Universitară Clujeană, Cluj – Napoca

Tafrali O (1936) Stațiunea preistorică din punctul Boghiu. Arta și Arheologia 11–12:51–54

Tarragüel AA, Krol B, van Westen C (2012) Analysing the possible impact of landslides and avalances on cultural heritage in Upper Svaneti, Georgia. J Cult Herit 13:453–461. doi:10.1016/j.culher.2012.01.012

Thiery Y, Malet J-P, Sterlacchini S, Puissant A, Maquaire O (2007) Landslide susceptibility assessment by bivariate methods at large scales: application to a complex mountainous environment. Geomorphology 92:38–59. doi:10.1016/j.geomorph.2007.02.020

Văleanu MC (2003) Omul și mediul natural în neo-eneoliticul din Moldova. Editura Helios, Iași

Wachal D, Hudak P (2000) Mapping landslide susceptibility in Travis County, Texas, USA. GeoJournal 51:245–253

Conclusions

The present study, with a pronounced interdisciplinary character, based on a small territory (which can rightfully be considered the cradle of Neolithic civilisation from Eastern Europe), has reached a sensitive point in the research focusing on environmental factors, in a tight correlation with the placement and distribution of archaeological sites. One of the main advantages of studying small catchments is that the results can be extrapolated for larger catchments (>100 km^2), the volume of analysed data being smaller, but done with the same scrutiny. During the three years of doctoral studies, a significant part of the observations made were in the field, which is represented as being one of the most important parts of this study.

All the obtained information through the GIS analysis was stored in digital format. The settlements were individually mapped, with high precision, with the use of geodetic GPS: during the field research done along with archaeologists from Romania and from abroad, some imprecise or insufficient descriptions were corrected and/or completed. If, at the beginning of the research a number of 23 Chalcolithic sites were known, presently their numbers have risen to 26, being discovered, mapped and partially dated: *SV de Boghiu* – Cucuteni A (Filiaşi, Bălţaţi commune), *Dealul Harbuzăriei/vest de Boghiu* – Cucuteni unknown (Filiaşi, Bălţaţi commune) and *Dealul Hârtopului* – Cucuteni A–B (Boureni, Târgu Frumos).

In every chapter which describes the natural framework, the archaeological factor is also analysed and completed with conclusions regarding the placement and evolution of sites in different natural conditions. This is where the GIS helped in obtaining the morphological parameters and afferent statistics (hypsometry/site placement on hypsometrical classes, slopes/site placement on slope classes). The predisposition for placing the settlements close to a water course or spring was highlighted, also on terraces and structural plateaus as visibility was better.

For the upper part of the basin, where the topographic plans were missing, measurements were taken continuously over a period of a week with the help of GPS (RTK mode) and colleagues from the Interdisciplinary Research Platform Arheoinvest. The obtained data was integrated and processed with the help of GIS. As a result, a DEM was developed for the upper basin, as no digital information was available for 1:5000 scale and has been found to be extremely useful for the application of geographical methods in archaeology (the DTM obtained after

© The Author(s) 2016
I.C. Nicu, *Hydrogeomorphic Risk Analysis Affecting Chalcolithic Archaeological Sites from Valea Oii (Bahlui) Watershed, Northeastern Romania*,
SpringerBriefs in Earth System Sciences, DOI 10.1007/978-3-319-25709-9

digitisation of topographical maps at a scale of 1:25000 was found insufficiently detailed).

Analysing the natural hazards (landslides, gully erosion, backwater) that have a direct effect on present-day populations makes this field of study a highly resear-ched subject on a global level, on the basis of climatic global changes and of anthropic interventions; this approach was applied for the archaeological sites in the hope of protecting the cultural heritage, and becoming a novelty on a national level.

Field mapping of all the hydrogeomorphological processes combined with the utilisation of high resolution satellite images resulted in a detailed cartographic background of the entire basin, all of which presently affect the archaeological sites due to hydrogeomorphological processes in different stages of evolution. The anthropic processes were taken into consideration where possible. The importance of protecting and conserving the cultural heritage must be underlined, being an essential part in preserving the national identity and salvaging the cultural heritage for future generations. The results of this study led us to obtain and salvage important information about archaeological heritage.

The three case studies analysed (*Dealul Mănăstirii/la Dobrin/Dealul Gosanul, Dealul Boghiu/Dealul Mare/Filiași, la Iaz/Iazul 3/Dealul Mândra*) are typical examples of archaeological sites which are under the direct effect of soil erosion processes, incompetent management disregarding the regulations for heritage pro-tection, and anthropic interventions. These studies have shown a series of research irregularities, resulting in inadequate information obtained for this field of study and the insufficient action taken by local authorities or certain private businesses. The lack of collaboration between the geographers and territory planners is obvious, with the areas becoming degradated and affecting archaeological heritage over time. The proposed land improvement measures are made at an empirical level and could constitute a real basis of solutions and approaches which would serve local authorities when starting land improvement works.

Another aspect is the touristic potential of the existing sites. At Cucuteni there is a museum dedicated to this civilisation; visitor numbers have been decreasing every year, due to the unsatisfactory promotions from the local authorities. The results of this study might be a valuable one in increasing the material and information for this museum, forcing the local authorities to promote and raise the public interest for the Chalcolithic civilisation.

Some locations that were picked for settlements within the basin had good visibility between them; we can conclude that Chalcolithic populations had a good understanding of what was happening inside the basin. This fact is not decisive in the placement of all the settlements, considering the numerous factors with a complementary character (proximity to water sources, terrains adequate for agri-culture, etc).

Revealing existing links between man and environment among these civilisa-tions, on the premises of actual evidence left behind by these cultures, constitutes the backbone of this thesis. Some models of migration and abandonment were highlighted for the entire basin, information was obtained based on scientific archaeological facts; the international research literature was applied for this study on a national level—the Island Biogeography, which indicated the same migration tendency from the middle to the upper basin beginning with the second stage (Cucuteni A–B), based on the Holocene climatic variations. Another important point within this research is represented by the analysis undertaken of the salted areas (soils with a high concentration of salts, not of salted streams) closely connected with the placement of archaeological sites, and also their economic importance for the current population.

Even if the great civilisations of the world were developed and placed on low alluvial valleys (Nile) or in coastal areas with high productivity (Peru), the Chalcolithic population has preferred the higher areas with adequate relief for better protection.

Printed in the United States
By Bookmasters